MEANDERING IN MEDICAL PHYSICS

Other history of science titles from Institute of Physics Publishing include

R F Mould: *A Century of X-Rays and Radioactivity in Medicine*
'... illustrated richly with an encyclopaedic collection of most curios and enlightening prints depicting the entire course of the development of radiology in medical practice from the time of the discovery of X-rays and radioactivity from 1895 onward.'

International Journal of Radiation Oncology Biology

S Webb: *From the Watching of Shadows*
'To quote from Isaac Newton, "the modern practitioner stands 'on the shoulders of giants'"... an excellent source of reference material.'

HPA Bulletin

A J Meadows and M M Hancock-Beaulieu: *Front Page Physics*
'... very different from the usual run of popular science books... Fascinating to read first hand about "the mystery of radium", the triumph of the general theory of relativity, or the first observations of the cosmic background radiation.'

The Good Book Guide

Twentieth Century Physics
'... ought to be in every library used by graduate or undergraduate students.'

Times Higher Educational Supplement

MEANDERING IN MEDICAL PHYSICS

A personal account of hospital physics

J E ROBERTS

(deceased)
BSc, DSc, PhD, Honorary Fellow of the IPEM, FInstP,
former President of the HPA, former President of the BIR

Emeritus Professor of Physics Applied to Medicine
University of London at Middlesex Hospital Medical School

And with an introduction by N G Trott

Institute of Physics Publishing
Bristol and Philadelphia

British Library Cataloguing-in-Publication Data

A catalogue record for this book is available from the British Library.

ISBN 0 7503 0494 4

Library of Congress Cataloging-in-Publication Data are available

Published by Institute of Physics Publishing, wholly owned by The Institute of Physics, London

Institute of Physics Publishing, Dirac House, Temple Back, Bristol BS1 6BE, UK

US Office: Institute of Physics Publishing, The Public Ledger Building, Suite 1035, 150 South Independence Mall West, Philadelphia, PA 19106, USA

Typeset in TEX using the IOP Bookmaker Macros
Printed in the UK by J W Arrowsmith Ltd, Bristol

To my wife,
who endured all this for nearly 64 years
and never complained

Professor J E Roberts, on the occasion of his 90th Birthday Seminar

CONTENTS

INTRODUCTION

Professor John Eric Roberts was born in the West Ridings of Yorkshire on 5 September 1907. He died on 14 October 1998 after a short illness. He married his wife Sarah in 1932 and was widowed in 1996. He is survived by his two daughters Sally and Wendy. He lived a full life in science and in his Methodist Christian faith and made friends all over the world. Obituaries in the national press, *The Independent* (5 November 1998) and the *Gazette* (19 November), present fine records of his achievements. Many of his former colleagues came to his 90th Birthday Celebration held in Bristol on 19 September 1997, organised by Professor Angela Newing of the Physics Department, Gloucester Royal Hospital. He clearly enjoyed meeting so many former colleagues, and the entertaining talks given in the accompanying seminar.

After graduating and gaining a PhD in the Physics Department at Leeds University, Eric Roberts worked with W V Mayneord from 1932 to 1937 at The Cancer Hospital (Free) in South Kensington, investigating the dosimetry of X-rays generated at 400 kV. He then took up a post at the Medical School of Middlesex Hospital. He was awarded the DSc for his original work in dosimetry and, following the retirement of Professor Sidney Russ in 1946, was appointed the Joel Professor of Physics Applied to Medicine. In his following 25 years at Middlesex Hospital Medical School, with its special links with the Hospital, he encouraged the advance of physics in varied aspects of hospital work; he was especially pleased with his success in stimulating the formation in 1961 of the Institute of Nuclear Medicine. From 1938 onwards he trained many young physicists, who later moved on to head departments of their own and became well known for their own contributions to medical physics. In 1943, 53 of them came together to form the Hospital Physicists' Association. In addition to his advisory role to the UK Government on matters involving physics and medicine, he was invited by the International Atomic Energy Agency to take part in training programmes in many parts of the world.

This book, to which he devoted so much effort, is, I believe, a unique account of a period between the mid-1930s and the mid-1970s: it graphically depicts the change between a world of home-made apparatus, hot valves and travel by steam train and ocean liner to the arrival of instantaneous communication throughout the world; the excitement of the first transatlantic flights to the present-day routine of jets. Professor Roberts made full use of all the available methods of travel to take his knowledge and skills to aid the advance of 'physics as applied to medicine' in many parts of the world. In his book he describes, in detail, some of his experiences in visiting and working in some 24 countries in the Middle and Far East as well as the USA and other countries.

I had various links with Professor Roberts over the years. We worked together on the first two revisions of the original Department of Health and Social Security Codes of Practice for safety from radiation hazards in the Health Service in the 1960s—under the outstanding Chairmanship of Sir Brian Windeyer. I succeeded him as the 'Non-medical' Editor of the *British Journal of Radiology* in 1962 and we were both Presidents of the British Institute of Radiology (though I was elected 28 years later!). But, in addition, when in the 1980s I took on responsibility for the BIR Archives, he deposited interesting historical material in these archives and so it came about that I had a letter from him in late June 1998 asking for help in providing these and some other archive material for use as illustrations in this book. Mr J E (Bob) Burns had just taken over the role of BIR Honorary Librarian and he with others including Dr Alan Jennings and Mr D R Guttery took this task on. We were very sad that Professor Roberts died when the work was still in progress but with help from others, including Professor Frank Farmer (who recalls with pleasure the time he worked with Eric Roberts at the Middlesex Hospital some 50 years ago) and the BIR Librarian Mrs Kate Sanders, the project was completed.

Nigel G Trott BSc, PhD, CPhys, FInstP
Emeritus Reader in Physics as Applied to Medicine,
Institute of Cancer Research, University of London

PREFACE

It is probably desirable to make clear that in the meanderings of this little book there is no intention of writing a formal or definitive early history of medical physics, either in Britain or elsewhere. The first 40 years have, in fact, been effectively covered by the 40th anniversary volume produced by the Hospital Physicists' Association, now the Institute of Physics and Engineering in Medicine. I can only suggest that any history in these pages is neither formal nor definitive but rather personal and anecdotal. My picture is one of the aged professor, comfortable in the Common Room, yarning about the old days in hospitals and laboratories, with a range of anecdotes, some of which, perhaps, had little or nothing to do with medical physics. I trust most readers will accept this innocent interpretation.

Preliminary reading of the work suggested that the meandering was more like hopping about in time and space. In the interests of some degree of coherence, I have divided the travels into two parts. The first part is largely concerned with meandering in time through the development of medical physics, mainly confined to my own country. In the second part I have almost literally meandered around the world, observing and occasionally giving a little help to the spread of our profession in many places, some of them perhaps rather unexpected.

When I first walked through the doors of The Cancer Hospital (Free) in 1932, I doubt whether the words medical and physics had very often been put together. It is still a relative newcomer to the scientific vocabulary but now well accepted and respected. Medical physics has been at the centre of my life and I am most grateful for a remarkable lifetime experience. It is usual, I know, in a preface, to thank people, but there are far too many here deserving of such thanks. They are all there, somewhere in these pages, and to every one of them I offer my sincere thanks.

JER
Badgeworth
August 1998

LIST OF ILLUSTRATIONS

CHAPTER 1

A NEW PROFESSION

Medical physics or hospital physics or, to use the modern up-market designation, physical sciences in medicine, must be regarded largely as a twentieth-century profession or phenomenon, although it is possible to trace links between physics and medicine over long periods of history. Over the course of recorded history, there has been little separation of the sciences, and until recently, the medical man (or should I say 'medicine man'?) has been regarded as being wise in many other fields, including science. From the Middle Ages onwards, as the sciences slowly separated out, the overall weight of knowledge remained such that one intelligent person could cover more than one field. From time to time there were individual medical practitioners who dabbled in some aspects of physics, though the two subjects never came together academically. Having written all this, I have to record two modern exceptions within my own acquaintance, thus illustrating the 'meandering' nature of these notes. Harry Flint (1890–1971) graduated in physics and reached the highest level of DSc whilst a reader in physics at King's College, London; then, still in post, he became a part-time medical student and qualified in both medicine and radiology, before moving to Bedford College as professor of physics. He continued to practise both physics and medical radiology. Edward Williams qualified in physics and came to do postgraduate work in medical physics at the Middlesex Hospital, but soon changed direction. He became a medical student and also acquired a PhD in physiology. He then came, via radiotherapy, to us in nuclear medicine and eventually succeeded me as Director of The Institute of Nuclear Medicine, becoming professor in 1971. These and a few other exceptions apart, it is interesting to note that, in earlier days, so far as we can tell, those few medical practitioners who dabbled in physics rarely, if ever, strove to link the results of their experimental physics investigations to their clinical practice. In other words, until quite recently there was no clear demand for the multidisciplinary subject of medical physics nor for the associated profession. Nevertheless, from the late nineteenth century onwards, a knowledge of basic elementary physics was deemed essential for any student who aspired to medical qualifications. I feel sure that many of my older colleagues who suffered from 1st MB (1st exam for the Bachelor of Medicine) physics would express very mixed feelings on the subject

It should be pointed out, however, that, with some exceptions, the physics taught in those courses could in no way be described as medical physics. Most of it was very basic, with little attempt at applications, medical or otherwise, part of education rather than professional training. It is, perhaps, a little strange that, at a time when medical physics is still an expanding profession and when the advances of physics into medical practice are quite phenomenal, the teaching of physics to medical students has virtually disappeared. In my own teaching days, we managed, with the connivance of physiologists and others, to insert a little more advanced applied physics into the pre-medical curriculum. More important, though, was the fact that the clinicians were encountering more and more physics-based equipment, techniques and principles in their everyday experience. Though there may have been mourning or delight at the disappearance of physics from the medical curriculum, the fact has to be faced that, over the past 50 to 100 years, the physical sciences, with their associated technologies, have effectively transformed the face of clinical medicine. Many brave clinicians have coped with and enhanced the new 'high-tech' modalities, as can be seen in the specialist journals, with a high proportion of papers presented jointly by medically qualified and non-medical authors. More than this has been required and in the past 60 years or so there has grown up at least one new profession to help deal with this revolution in medical practice. It is normally designated hospital or medical physics, though in some areas, with a slightly different emphasis, we also have biological engineering.

1.1 Röntgen and the Discovery of X-rays

The year 1895 might be regarded as a turning point in history; it was undoubtedly a turning point in medical history and probably in that of physics. We might, then, regard it as the point of emergence of our new profession of medical physics. As is well known, in that year, the existence of X-rays and some of their strange properties were

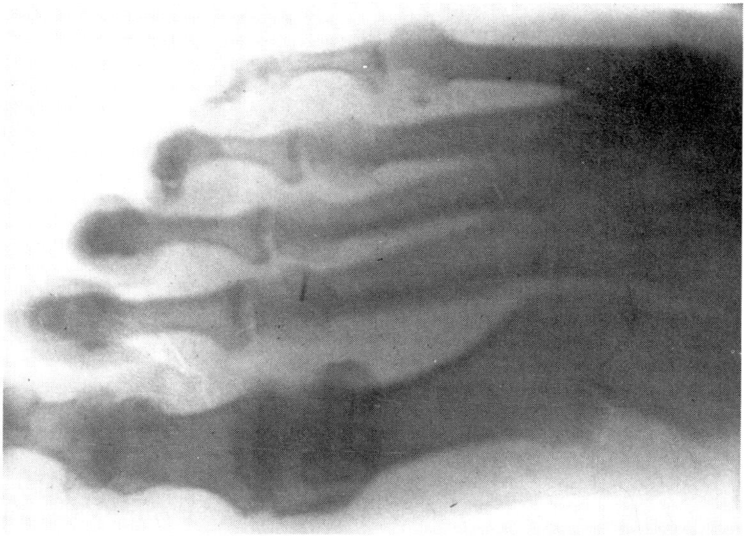

Early radiograph of a human foot with embedded needle.

demonstrated. The discovery was made in the relatively undistinguished physics laboratories of the University of Würzburg, in Germany, by a modest professor of physics, Wilhelm Konrad Röntgen. In itself, the discovery had no medical connotations whatever; medical physics can make no claim. Röntgen was trying to make a contribution to the rapidly expanding physical study of atomic structure, in which electric discharge in gases was then a most important tool. Eventually, X-rays would be seen as a very significant factor in the understanding of the structure of atoms and later as a tool in the investigation of the structure of crystals and other forms of solid matter. To contemporary physicists it opened up some new and exciting fields which required detailed follow-up. For a few discerning clinicians who read of the discovery, it was dynamite. Professor Röntgen had seen the details of the bones of his wife's hand from the outside without surgical interference. It is indeed difficult, surrounded as we are now by the immense range and sophistication of modern radiology, for us to appreciate just what that meant. In the Dean's office at the Middlesex Hospital Medical School I often gazed at a radiographic print showing a human hand with a needle embedded in it, the first radiograph made at the Middlesex in 1896. The succeeding radiology by which I was surrounded some 60 years ago would be regarded by today's standards as primitive, but that print filled me with awe and wonder.

Röntgen's dramatic discovery started a tremendous surge of activity in the medical circles of most developed countries. In Britain, the Rontgen Society was founded in 1897 and its first president was a physicist, Professor Sylvanus Thompson; it might be noted that Sylvanus Thompson could, in no way, be regarded as a hospital or medical physicist. That particular genre did not make its appearance until some years later. There is an interesting comment on the speed with which some aspects of physics penetrated medicine in Professor Thompson's first Presidential Address. He refers to such things as fluorescent and intensifying screens, tube regulation, special films for X-rays, cineradiography, localisation and dermatitis—all this within two years of the discovery of X-rays. From the start, the Rontgen Society and its eventual replacement, The British Institute of Radiology, accepted both medical and non-medical members. Now, having achieved its centenary in 1997, it continues to provide a meeting ground for all the professions involved in medical radiology. It could be said that the seed which eventually produced the emerging of the hospital physicist germinated in these surroundings.

In the light of the previous observations, it is interesting to note that in the early days following the advent of X-rays, the hospitals and medical practitioners rarely recruited physicists to build and maintain their radiological units and departments. Among the general public there was clearly some suspicion about the new rays; some people did not take kindly to the idea that doctors and others might see what went on inside their bodies. It may be a consequence of this attitude that in the first quarter of this century, the common practice was to install the dreaded X-ray equipment in basements, often remote and inaccessible. For the design, construction and maintenance of any apparatus which could produce X-rays, hospital engineers, pharmacists (a surprising but common choice) and other people who would nowadays be designated handymen or DIY enthusiasts were recruited. It would seem that for the first 30 years, with one or two exceptions, professional physicists, who were academically interested in the properties and technology of X-rays, had very little influence on their clinical applications. Fortunately for the future of medical radiology and, indeed, of hospital physics, by the end of the First World War, a substantial X-ray industry was being established.

Even in 1932, when I myself entered the field, the very idea of a graduate physicist being appointed directly to a hospital post was considered rather peculiar by some people, particularly in the universities. I am sure most of my colleagues in Leeds University (except one, F W Spiers, who soon followed the same track) thought me just a little mad when I announced I was going to The Cancer Hospital in London, instead of to a safe lecturing job in another university. However, in the late 1920s and early 1930s, things were changing. By that time the days of the amateur and the DIY enthusiast in the X-ray department were coming to an end. The gas tubes and the glowing 'modern' Coolidge tubes in lead-lined boxes were disappearing. Sophistication in equipment was becoming the name of the game. One important factor was that a small number of large industrial organisations were beginning to see profitable possibilities in the market for X-ray equipment. Possibly as a matter of cause and effect, a small number of new physicists began to be appointed directly to hospitals. Within a very few years of my own arrival, Bill Spiers had started up in Leeds, Herbert Parker took on both Manchester and Ralston Paterson, Harold Gray appeared at Mount Vernon, L H Clarke moved from the Middlesex Medical School to Lambeth and Joan Honeyburn, the first lady, joined us at The Cancer Hospital in the Fulham Road. In the light of the later development of hospital physics, these all appear to be significant moves. Added to this, the physics department staffs at a few medical schools, particularly Bart's, Middlesex and Guy's in London, though not strictly hospital appointees, were making increasing contributions in the clinical field. Up to 1939 and beyond, with a few minor exceptions, the work of a physicist in a hospital was confined almost entirely to the applications of X-rays and radium in the radiotherapy department.

1.2 Medical Physics in Britain in the War Years (1939–1945)

During the war years of 1939 to 1945 there was an inevitable slowing down in this expansion of hospital physics, but some progress was still made in the development of radiological applications of physics. A number of important technical innovations in radiological equipment were already in place. Naked high-voltage leads were disappearing and enclosed high-tension generators with sealed X-ray tubes were in regular use. Bart's even had Britain's first million volt installation, which was not shock proof. In a number of places radiotherapy was maintained by the installation of the very robust Metropolitan Vickers continuously evacuated tubes and generators. These, though a relatively short-lived development, proved a most valuable link between the fairly primitive therapy equipment of pre-war years and the dramatic developments in supervoltage therapy of more recent times.

All this led to a further recruitment of physicists in the hospitals and quite a number of subsequently distinguished founder members of the Hospital Physicists' Association (HPA) appeared at this time. It is really rather surprising that right in the middle of a major war in 1943, the HPA should be founded, largely on the initiative of Professor Sidney Russ. The pioneers of the then new profession in Britain in 1943 numbered just 53. This is no place to record the history of the HPA which, in any case, has been adequately done elsewhere. Suffice it to say that, as these notes are being written, the Institute of Physical Sciences in Medicine, scientific arm of the old HPA, has combined with the Biological Engineering Society to form the Institute of Physics and Engineering in Medicine, with well over 2000 members. There are now similar national groups of medical physicists in many developed countries, linked to continental groupings in Europe, America and Australia, all being bound up to the worldwide International Organisation for Medical Physics. All this represents a dramatic explosion in expertise, interest and professional standing from our little group of 10 or so in 1932 and the pioneering 53 membership of the HPA in 1943.

1.3 Developments in Radiology and Radiotherapy

When the war was over there was an equally dramatic explosion of new physical knowledge, technical development and practices. So far as medicine was concerned, it was not surprising that this started in the fields of radiology and radiotherapy. The modernisation of the old X-ray departments which began in the late 1930s was soon completed. No longer were the radiologists, the radiographers and their patients in danger of electrocution in their X-ray rooms. For diagnostic radiology, high-load rotating-anode tubes became commonplace and such things as tomography and similar elegant techniques appeared. In the high-voltage generators, the glow of thermionic rectifying valves disappeared almost overnight; wartime solid state physicists had produced more effective alternatives. In radiotherapy, 200–250 kV X-rays were effectively standardised and moves were quietly afoot towards millions of electron volts. Many new developments from wartime military research were in the pipeline. For radiotherapy perhaps, the most important and dramatic came from radar. In a Birmingham University physics laboratory, the magnetron was produced—an elegant means of generating at high power the microwaves now so familiar in military installations, aircraft and kitchens. Combine these with a suitable waveguide and an appropriate magnetic field and we have an accelerator giving the radiotherapist 5, 10, 15 million electron volts or more. Still with radiotherapy, the researches which had produced the atomic bomb led directly to the advent of the sadly named 'cobalt bomb'. Nuclear reactors could produce large quantities of cobalt-60 and later caesium-137 and so in the 1950s telecurie units spread around the world at a tremendous rate.

There was a similar explosion of change in diagnostic radiology, with a whole range of tricks to enhance and analyse the radiographic image. The most dramatic of these developments arose through the application of what would now be regarded as somewhat primitive computer technology. This produced precise computerised tomographic images of thin sections of the anatomy and would be regarded by many people as the greatest advance in diagnostic radiology since the appearance of the image of Frau Röntgen's fingers.

1.4 Developments in Ultrasound and MRI

Still with medical imaging, just a little later, straight from the physics laboratories of the world came two almost incredible developments. For many years, ultrasound had been used in some fields of medicine but with no major impact. Then, suddenly, with improvements in ultrasound sources and computerised scanning techniques, we had ultrasound imaging of surprisingly high quality. This has proved very attractive, particularly in fields where there were possible health risks from ionising radiations. Now the ultrasound scan is widely used, especially in obstetrics and gynaecology procedures. To introduce a personal note, I have even seen a picture of my latest great-grandchild, some months before her birth; for me, this is an almost unbelievable late contact with my beloved medical physics. The second, even more dramatic, surprise came from a most abstruse field of experimental and theoretical physics. The academic physicists were interested in the resonances which arise in electrons and nuclear particles when matter is placed in an intense magnetic field and subjected to a very high-frequency electromagnetic field. These resonances depend on the molecular structure of the materials exposed in the field. By an enormous experimental extrapolation, there arose the possibility of displaying an image of the variations of these resonances in a heterogeneous sample of materials and eventually in a section of the human body. Thus, very crudely explained, was born the new technique of nuclear magnetic resonance imaging (MRI). I sincerely hope the exponents of this highly sophisticated technique will forgive my grossly oversimplified description. The fact remains that, after some 80 years of diagnostic radiology with X-rays, within a very few years we had two entirely new methods of imaging human anatomy, with not a single radiation-induced mutation in sight.

1.5 Developments in Nuclear Medicine

With all this going on in the various branches of radiology, it was inevitable that the size and influence of the physics departments should steadily increase. It was just as inevitable that their influence should spread far beyond radiology. Indeed, the ideas and products of physics and its associated engineering have invaded almost every branch of medicine and surgery. Perhaps the most extensive invasion, again associated with radiation, involved the birth of a new medical speciality, nuclear medicine. Not long after the Second World War, small quantities of radioactive nuclides became available from experimental nuclear reactors. This gave rise to endless thought and speculation on the possible uses of these isotopes in the investigation of chemical, biological and metabolic processes. The physicists in the hospitals had much of the know-how and most of the necessary equipment and it was not very difficult to tempt enterprising clinicians into the field. It just happened that one of the most readily available isotopes was iodine-131, almost made to measure for the study of thyroid function. It is a sobering thought that the birth of a whole new medical speciality depended on the success of those rather crude early studies of iodine metabolism. In the event, radioactive nuclide studies have become highly sophisticated and have invaded large areas of medicine, surgery and radiology. Some hospital physics departments have retained their nuclear medicine units but in other cases new clinical departments and institutes have grown up to cover the subject. In more recent times, nuclear medicine has developed a range of sophisticated imaging techniques and this has produced links with conventional radiological imaging to lead to the discipline of clinical radiology.

1.6 Developments in Physics and Associated Technologies in Other Areas of Medicine

Since the Second World War, in addition to advancing the technology of radiology and giving birth to nuclear medicine, physics and its associated technologies have spread far and wide through medicine and surgery. One has only to look around at the intensive care unit of any major hospital, an up-to-date operating theatre or, say, a cardiology unit to see a range of scientific instrumentation which would have been inconceivable in a university or industrial physics laboratory 50 or 60 years ago. Developments in the optics of endoscopy and the miniaturisation of many instruments are beginning to transform many aspects of surgery, with an increasing range of procedures being carried out without incisions. When I was first introduced to lasers, they were of great academic scientific interest but were usually demonstrated as laboratory conjuring tricks. In one direction they soon became an essential part of the illumination of 'rock' music performances and discos. They have now become an important part of the surgeon's armoury. In the fields of rehabilitation, prosthetics and the like, teams of physical scientists and engineers have grown up and, in collaboration with the clinicians, have almost revolutionised the situation. Some people would regard the computer as the brightest example of the advance of the physical sciences in recent years. One thing which strikes an ageing hospital physicist when he now visits hospital departments is that he finds some form of computer terminal in almost every office, laboratory and ward. Only very recently, I was visiting our local, small 'cottage' hospital and there, at the geriatric ward nurse station, was a very substantial computer terminal, with Sister working hard on it. Almost all of this development has taken place in the course of one man's retirement. Even so, this little attempt at history does less than justice to the impact of the physical sciences on clinical medicine over the past 50 years. It does, however, offer some explanation as to why the number of physicists and engineers employed in the medical services has increased so rapidly over recent years; in effect, it explains the growth of that new profession of hospital physicist. It all appears to have come from a tiny handful of physicists in Europe and America who, in the late 1920s and early 1930s, saw interesting problems in the use and measurement of X-rays and radioactivity. Some of them even had the impertinence to think they might be of assistance in the applications of the new radiations to clinical problems. Now, hospital physicists and bioengineers are numbered in thousands throughout the world and the range and sophistication of their activities is almost breathtaking.

1.7 Conclusions

What has been written in this opening chapter can in no way be regarded as a formal, definitive or researched history of the profession. What I have tried to do is to sketch a background picture against which I meandered for over 50 years. The emphasis is quite deliberately on the earlier days, on the assumption that the activists of today are familiar with more recent developments. These meanderings have followed the whole range of social and professional life, with little emphasis on scientific tricks or technical success. Hospital physics has been very good to me and it has all been an incredible experience. Retirement inevitably brought a slowing down and eventually an end to the main meanderings. For nearly 30 years now, I have been able to meander along the sidelines and watch, with continuing amazement, what the activities of physical scientists and engineers have continued to do to the practice of medicine. The physics departments of our two nearby general hospitals, Gloucester and Cheltenham, welcome me regularly and though I think they regard me as something of a mascot, they blind me with their science and I still feel part of the process.

CHAPTER 2

SETTING OUT

2.1 The University of Leeds

My very first contact with medical physics occurred some time in 1930 and that contact illustrates beautifully the very primitive or almost non-existent state of the art some 65 years ago. At the time I was a humble postdoctoral research assistant in the University of Leeds, existing on a grant and nominally helping the professor with the continuation of the research work I had done for PhD. This was the standard recipe for a career academic physicist. One day there was an urgent call to the Physics Department from the Huddersfield Infirmary. They had lost a number of radium needles from the operating theatre and had no facilities for searching for such things. Could the University Physics Department help them? For myself, I had not the remotest idea what a radium needle looked like and I am fairly sure that the same could be said for the rest of the Department staff, though they regularly taught the students all they knew about the properties of alpha, beta and gamma rays. This was clearly a job for the most junior member of the academic staff, and he could easily be spared, so I was duly nominated.

The equipment which, on the advice of my more senior physicist colleagues, I took with me would only raise howls of mirth today. The main item was a 10 cm cube gold leaf electroscope with a home-made sulphur insulating plug, straight from the First Year Teaching Laboratory. To charge the electroscope, I was supplied with an ebonite rod and, quite literally, a cat's skin. From such beginnings has arisen the whole modern range of sophisticated instrumentation for the detection and measurement of ionising radiations. With this baggage I was off by train to Huddersfield (in those days, cars were limited to professors) and for two long days I searched the operating theatre, laundry and drains. False alarms were frequent as the humidity caused the electroscope

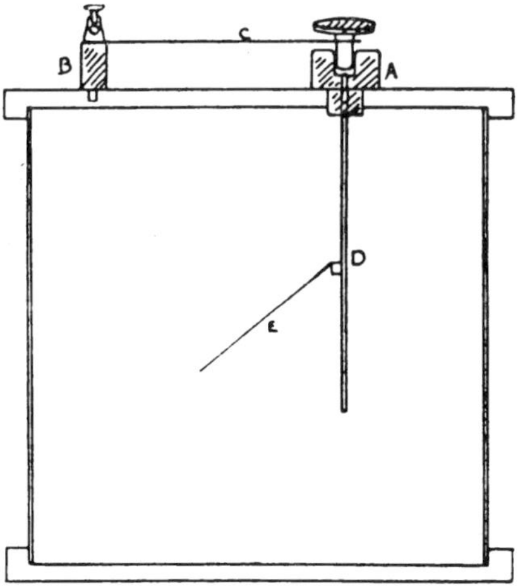

A. Amber block
B. Ebonite pillar
C. Fine steel spring
D. Elektron rod
E. Carbon fibre

Gold leaf electroscope, reproduced from Mayneord 1929 The Physics of X-Ray Therapy (J & A Churchill).

insulation to leak and give false indications, but, sad to say, the radium was never found. It was a fascinating, if inelegant, introduction to hospital physics life. At that time, little did I imagine that I would spend practically the whole of my professional life applying my hard-learned physics to medicine and hospital problems.

2.2 A First Visit to The Cancer Hospital (Free)

By 1932, I was ready to move on, almost certainly on to the first rung of the conventional academic ladder, as a demonstrator or junior lecturer in another university. Indeed such a post was in the pipeline for me. Then I saw an advertisement for a graduate physicist of about my level at The Cancer Hospital (Free), in London. This was an extraordinary idea, which greatly intrigued me, and I discussed it with my senior colleagues. To my surprise, I discovered that one of them, F W Spiers, just senior to me as a demonstrator, had already applied for the post. I quickly decided to forget the whole idea but Bill Spiers himself persuaded me to apply, if only for a trip to London. In fact we travelled up together. After two days of intensive interviewing, to our amazement,

The Cancer Hospital (Free), 1934, courtesy of The Royal Marsden NHS Trust.

I was offered the post of Assistant Physicist at the munificent salary of £350 per annum, which was £50 more than I could have expected anywhere else. I learned later that one member of the Hospital Board, who shall remain nameless as he was a physicist, objected strongly to the inflated salary, insisting that he could get any number of PhD physicists for £250 per annum. One small event in that first visit to The Cancer Hospital remains firmly in my mind. In the Research Institute I was shown with tremendous pride a test tube containing a few grams of bright yellow crystals. This was 1.2.5.6 dibenzanthracene, the first pure specimen of a chemical carcinogen to be isolated and characterised. What really impressed it on my memory was probably the fact that the isolation and characterisation of the compound was greatly assisted by the spectroscopy carried out by Mayneord in the Physics Department. Although we rarely used it, the big Hilger spectrometer stood in a laboratory in the Department to remind us of this triumph. My outlook was already being broadened and I was seeing a very different approach

Professor Sidney Russ, President of the Rontgen Society (1919–1920), courtesy of the British Institute of Radiology.

to physics research, as compared with that of an academic department. Perhaps I should add the sad and completely irrelevant story of the end of our visit to London. Bill Spiers and I had made a deal that

Professor W Valentine Mayneord, President of the British Institute of Radiology (1942–1943), courtesy of the British Institute of Radiology.

if either of us was offered the post, that person would pay for the train dinner on the return journey. I had thought I was very safe. Unfortunately, one consequence of me being offered the post was that I was not, in accordance with the practice of the time, offered my travelling expenses. I remember clearly that payment for the dinner took every penny I possessed. I should add yet another postscript to this meandering story. It would seem very probable that my success in London was determined by some much higher authority with a special interest in my native city, Leeds. Only a year or so after these events, Bill Spiers was appointed Physicist to The Leeds General Infirmary and later to a Chair in the University, where he became one of the great medical physicists of our generation.

The die was cast. In writing this short section, I have strayed from serious consideration of medical physics to illustrate the nature of that die-casting process. There would seem to be a distinct sense of meandering into unknown territory and, at the time, I am sure that description would have expressed my feelings very well indeed. Sixty years or more later, I still meander into my neighbouring medical physics departments at Cheltenham and Gloucester, to be welcomed and vaguely recognised as one of the old boys who started it all.

2.3 The First True Hospital Physicists

In 1932 I am quite sure I did not realise that I had joined a very elite company, a group of about 10 or a dozen people who were the first true hospital physicists in the world. The doyen of them all was Professor Sidney Russ; though strictly based in the Medical School, he had held the post of Physicist to The Middlesex Hospital for almost 20 years. Val Mayneord had moved from St Bartholomew's Medical College to The Cancer Hospital (Free) in 1927. At the Radium Beam Research Unit in Central London was Leonard Grimmett, assuredly one of the characters of early medical physics. Although he made some valuable contributions to radiological physics, particularly in instrumentation, he would probably have claimed that his greatest contribution to human welfare and happiness was as a 'pop' musician and particularly as a pianist. He ended up as a physicist at the M D Anderson Hospital in Houston, Texas. Although not in the strictest sense hospital staff, Professor Frank Hopwood and Professor Gilbert Stead, at Guy's, with their physics colleagues, contributed to medical research and clinical work.

Birmingham had Frank Phillips as Radium Curator and Physicist. Phillips was also something of a legend, reputed to have devoted at least as much time to making a fortune on the Stock Exchange as to the study of dosage in radium therapy. In the United States, Gioacchino Failla and Edith Quimby were well established at The Memorial Hospital in New York, with Otto Glasser in Cleveland. In continental Europe, the massive figure of Rolf Sievert dominated the scene at Tre Radiumhemmet in Stockholm, whilst Walther Friefrich was raising the profile of radiation physics in Berlin. It was a small but, for the period, very mobile group and, within a couple of years, I had met them all. Even now, I take pride in counting myself as having been on the fringe of that company of pioneers of the profession. Mention should also be made of a number of the national standardising laboratories which, because of the urgent problems involved in the measurement of ionising radiations, worked in close contact with the active hospital physicists. At the National Physical Laboratory in Teddington, there was the great G W C Kaye (known to physicists worldwide, with Laby, for his *Tables of Physical and Chemical Constants*), and two of his regular assistants, George Bell and Walter Binks. It was well known that, at meetings such as the Council of the British Institute of Radiology, Bell or Binks would sit behind Kaye to provide prompts and information as required. In the United States, at the National Bureau of Standards, was Lauriston Taylor and in Berlin at the Physikalische und Technische Reichsanstalt was Heinrich Behnken. For their services to radiation dosimetry they surely should count at least as Honorary Founding Hospital Physicists.

2.4 Work at The Cancer Hospital (Free)

Turning back to The Cancer Hospital, where all this started for me, it is difficult nowadays when physicists abound in most large hospitals to appreciate the culture shock when a relatively naïve physics graduate was deposited into such a clinical environment. Apart from the dash to Huddersfield reported earlier, I cannot recollect ever having been inside a hospital before. True, I passed the doors of The Leeds General Infirmary almost every working day for seven years but had little idea of what went on inside. The Cancer Hospital (Free), limited as it was then to the Fulham Road site, was a relatively small hospital and it was not possible to get away from cancer and cancer patients, a real shock for someone unprepared by medical school or nursing school. I still realise that the sight of a children's ward full of cancer patients almost prematurely ended the career of one budding hospital physicist. I am not ashamed to say that the thought that there was a job to be done and I was considered able to do it was a great help in such circumstances. Following such ponderous thoughts comes a much lighter recollection which has little or nothing to do with medical physics but which might be worth recording. A stranger coming to live in London, I had to find a lodging and was fortunate in being able to rent a large attic near Marble Arch. My first address in London was therefore by far the most prestigious I was ever likely to have: Seymour Street, Portman Square. Incidentally, the rent for my attic was one guinea (£1.05) per week, with 5p for breakfast and $12\frac{1}{2}$p for dinner.

As is well known, The Cancer Hospital (Free) soon became Royal and in 1948, with the coming of the National Health Service and long after I had left, it lost its (Free) label, eventually becoming The Royal Marsden. Then, in 1932, as now, the hospital was regarded internationally as a centre of excellence both in clinical care and in more fundamental research. Attached to the hospital was the Research Institute, where, at that time, Kennaway and Cook were carrying out their fundamental studies on chemical carcinogenesis. So far as I recollect, the Institute had no physicist and, when the need arose, called on the Hospital Physics Department, as for example in the spectroscopic analysis of 1.2.5.6 dibenzanthracene, mentioned earlier. The Physics Department, as such, was an integral part of the Hospital; in fact, during my time there, it was a section of the Department of Radiology, under the direction of the redoubtable Professor Woodburn Morrison. It should be noted that in 1932, radiotherapy had not been separated from radiology. A signpost to that inevitable development in the near future was the presence of Miss Constance Wood as Radium Registrar, the nearest thing to a radiotherapist at that time. This hospital basis of the Department prompts an unusual recollection. One day, the Chairman of the Hospital Board had some kind of brainstorm and decided that his prestigious medical and scientific staff should be under tighter control. The order went out that every member of the staff, from senior surgeon to junior physicist and student nurse, must clock in their exact time of arrival and departure from the Hospital. Woodburn Morrison's instructions and comment are worth recording, to the effect—you must

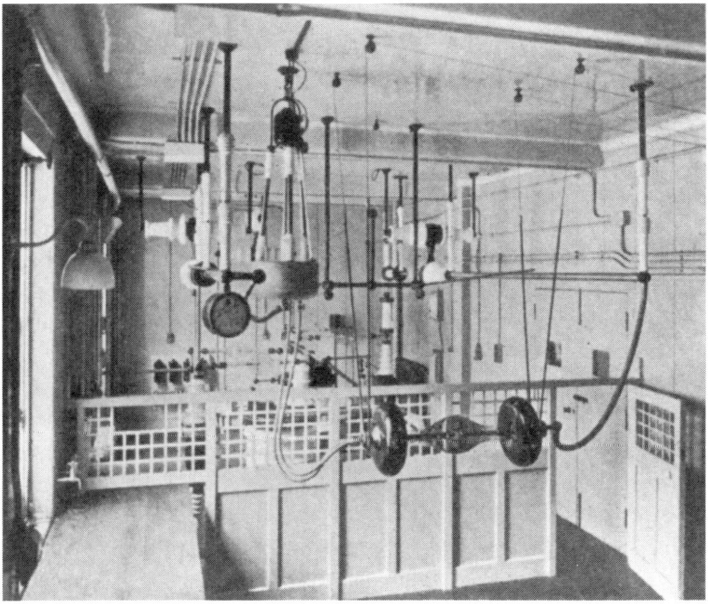

Siemens 400 kV X-ray machine, courtesy of the British Journal of Radiology, Volume 6, No 6, 1933.

clock in and out as ordered; what you do in between is nobody's business. The order was soon withdrawn.

By the standards of the time, The Cancer Hospital Physics Department was remarkably well equipped. It occupied one whole floor of the Radiology Block, some 20 per cent of the total space allocated to radiology and radiotherapy. The centrepiece was a large X-ray laboratory, in which had been installed what was probably the most advanced supervoltage (400 kV) X-ray machine in the world at the time. It had been designed and built by Siemens and contained all the advanced features of the X-ray therapy equipment of the period. Normal radiotherapy was just approaching the 200 kV mark; most sets were run at 180 kV, as they were rather unstable at the higher voltage. In our Physics Department machine, two 200 kV constant-potential generators, with tank transformers and brightly glowing thermionic rectifying valves (each about 3 feet long), had effectively been set back to back to produce 400 kV. A very large and very special tube of the exposed brightly shining Coolidge variety had been designed. No very serious attempt had been made to protect anyone from the very high voltages. The transformer tanks, capacitors, rectifying valves and high-tension leads stood in their nakedness behind a thin wooden lattice screen, through which came the naked connectors for the 400 kV tube. Before any adjustments to the gear could be made after switching off, it was necessary to lower an earthing bar to the output leads to discharge the capacitors. The control panel was in a small cubicle within the laboratory; it did, with some difficulty, house two people and was lined with some 10 mm of lead. In what would now be regarded as a rather naïve way, we were somewhat protection conscious. The British and International Protection Committees made recommendations on working practices but the only working dosage advice available at the time was that we should not work in any area where the ambient dose rate was greater than 10^{-5} röntgens per second. We did our best to follow this advice but I would be hard pressed to swear that we never consciously exceeded that limit. From time to time it was necessary to check the high-tension voltages on this and other installations and the only method then available was to use a calibrated spark gap. Consequently, I was drafted into the X-ray room to sit within 5 or 6 feet of the 400 kV X-ray tube to control the spark gap and take the voltage readings. Immediately above my head would be one of the high-tension leads, producing literally a hair-raising experience. What my radiation dose was on each occasion will never be known as we never tried to measure it. I have no doubt it would have led to apoplexy in any modern radiation protection adviser.

With all its virtues, the Physics Department's 400 kV set did present one or two technical and at least one social problem. As with many X-ray machines of the period, when working near maximum voltage there was a degree of instability, leading at times to dramatic displays of sparking in the high-tension system. The Siemens 400 kV was very prone to this effect; in particular it became unpredictable and sparked when movements of the Department's lifts disturbed the mains voltage. This made effectively impossible any series of precise ionisation measurements at the highest voltages during the busier parts of the day. Consequently much of our research work on ionisation chambers and on the extension of the use of the röntgen to higher energies had to be carried out in very unsocial hours between 8 pm and midnight. This did little for our social and family life but we could see no alternative. I still have a clear recollection of Mayneord and I sitting in the 400 kV control cubicle in the quiet of the night discussing the possible future of hospital physics. We had an idea that we might be in on the ground floor of something really big, a pretentious, even improper, thought, but we were not very far from the truth. Radiological physics, at least, had come to stay and many other undreamed of applications of physics were in the pipeline.

Returning to our survey of the Physics Department, there were two or three smaller research rooms, one of which was allocated to me as an office–laboratory. There was a well-equipped teaching laboratory which accommodated 10 or 12 students. One of the reasons for the increase in the physics staff, myself, was that the University of London had introduced its own Diploma in Medical Radiology in addition to the then existing Cambridge DMRE. These were, of course, later taken over by the Royal Colleges and eventually settled with the Royal College of Radiologists. It is most interesting to note how many of those embryo radiologists who passed through those laboratories in the early years were to become well-known names in the subsequent history of British radiology. To complete this brief description of that early Physics Department, we cannot forget the heart of it all, a very well-equipped mechanical workshop, under the control of the formidable Mr Henry Hodt. With this workshop, the standard of our in-house equipment and instrumentation was remarkably high. Furthermore, Mr Hodt, as should any self-respecting chief technician, ruled the staff, both technical and professional, with a rod of iron. It is nowadays interesting, almost incredible, to note that we had no electronics laboratory or workshop. Our limit, in what would now be just recognisable as electronics, was in the brightly shining rectifying valves in the X-ray generators and in our simple ‘wireless’ sets; no solid state rectifiers, no transistors, no silicon chips, no computer circuitry nor terminals. As we may see later on, somehow we overcame these deficiencies to some purpose. In total, with a comfortable study–office for the ‘chief’, we thought we were very fortunate in a most adequate department. It was small by present-day standards, though it seemed large for two graduate staff. The important fact was that we were undoubtedly on the way.

CHAPTER 3

EARLY DAYS

3.1 The Duties of a Medical Physicist

It is not difficult to see that the overall pattern and objectives of a modern medical physics department are not very different from what they were 60 years ago. The equipment has become increasingly sophisticated and expensive, the range of disciplines involved in physics and in medicine has extended enormously, but the three main aspects of a hospital physicist's duties have remained largely unchanged, namely clinical service, research and teaching. First and foremost must always be the provision of services and advice to clinicians and clinical departments for the care and welfare of patients. Over the years, some clinical departments have deemed it desirable to set up their own physics service sections, but in most cases hospital physics and engineering departments have, where necessary, subdivided their activities, whilst maintaining the advantages of centralised scientific control and particularly central technical services. In any physics department worthy of the name, there will always be a need to do some form of research work and to allocate time for this purpose. The research in a hospital department will tend to be of an 'applied' nature, designed to improve the services provided in clinical medicine. Over the years there have been many examples of such technical development, perhaps the most exciting being the three imaging projects mentioned in Chapter 1: computerised tomography (CT) scanning, ultrasound scanning and nuclear resonance imaging and spectroscopy, plus, of course, the wide range of nuclear medicine.

In our earlier context, these things were still a long way ahead, and it is not given to many laboratories to produce such miracles. Some hospital physicists have been fortunate enough to become involved in more fundamental research, though it might look difficult to find anything more fundamental, in this sense, than nuclear magnetic resonance. In the early 1930s, we were largely concerned with the most fundamental aspects of radiation interactions with matter and the resultant dosimetry. Thoughts were also turning to ideas on radiation interactions with matter at molecular and cellular level with new horizons in radiobiology.

The teaching of medical physics, apart from A-level style to first-year medical students, was originally limited to the Diplomas in Medical Radiology of the Universities of Cambridge and London and, for a time, Liverpool and Bristol. Somewhat later, hospital physicists were recruited to assist with the Diploma examinations of the Society (later College) of Radiographers. Formal teaching for the medical physics profession itself was a much later development, apart from appropriate postgraduate degrees, but with eventually a full training scheme for potential members of the Institute of Physics and Engineering in Medicine.

The service activities of medical physicists at the time when I joined the profession were very basic indeed. Being part of a relatively small hospital we were always in close contact with clinical matters, though there was not a great deal we could do for individual patients. We were occasionally asked to advise on the amount and distribution of radium for a surface applicator or an implant. It is perhaps difficult now, even for established hospital physicists, to realise that we had nothing like Paterson–Parker rules and tables to guide us. Even these were still a few years into the future. We were not even sure of that vital factor in radium dosimetry, the dose rate in röntgens per hour at 1 cm from a point radium source. About 1934, when really small ionisation chambers became available, Joan Honeyburn and I managed very laboriously to produce a definitive experimental picture of the distribution of dose near a radium surface applicator. In X-ray therapy, our main service commitment was the regular checking of the radiation output and quality on the treatment machines. Output, by then in röntgens per minute, was measured at least once a day and half-value layers in copper at regular intervals of a few weeks. For these purposes we were very fortunate in having several very 'modern' dosemeters of the Hammer or Mecapion type. These integrated charge from rather large thimble ionisation chambers and recorded doses in units of 5 röntgens, via an electromagnetic switch. Rather surprisingly, even at that early date, National Physical Laboratory calibrations of

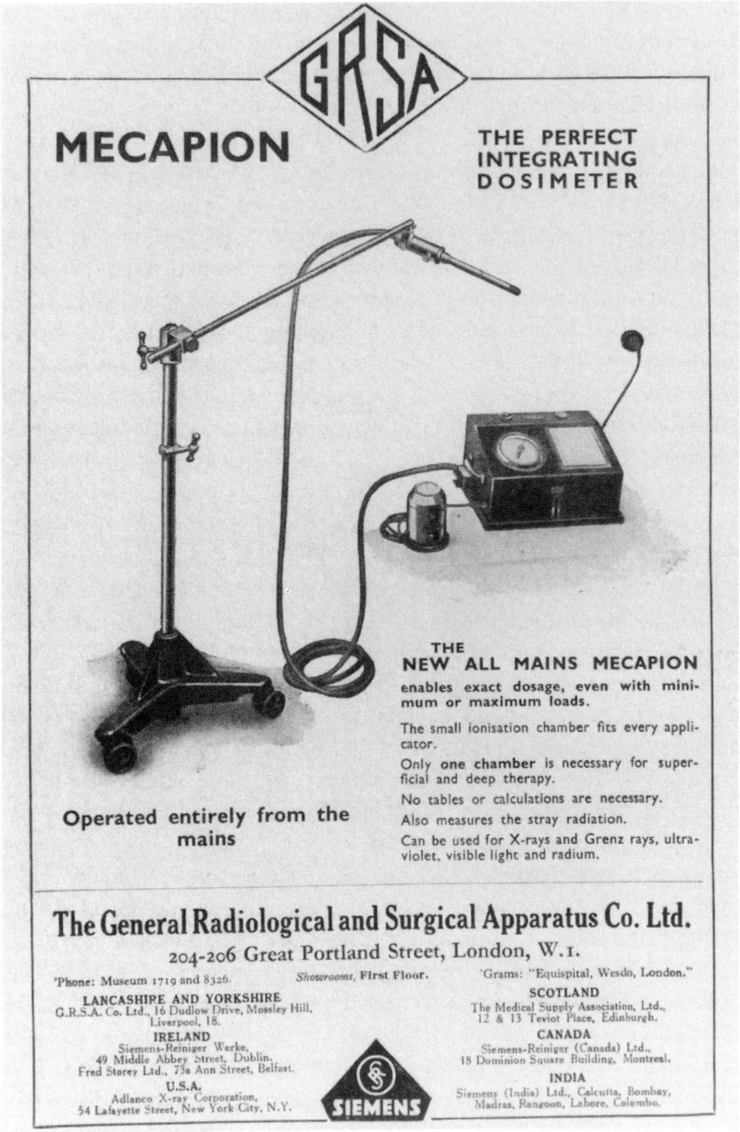

Example of old Mecapion dosemeter, courtesy of Dr R F Mould.

the dosemeters were available. Occasionally it was necessary to check the high-tension voltage on the therapy machines, using a spark gap as described in Chapter 2. This did no good at all to the accumulated radiation dose of the junior physicist. Most 'deep' X-ray therapy was carried out at about 180 kV. Those were the days when it was not unheard of for a patient to exclaim to doctor or radiographer, 'that big spark during the treatment did me a world of good!' At The Cancer Hospital in the early 1930s we were very fortunate in having, in addition to the Physics Department's experimental high-voltage machine, a 400 kV machine for therapy, designed and built by Ferranti. I believe this was Ferranti's only venture into this field and it had the then standard exposed high-tension equipment, valves and tube. I cannot resist a small anecdote from the history of this set. I was in the control room as a patient was being treated when, suddenly, the machine cut out completely. We soon realised that this had been caused by the opening of the corridor door to the treatment room and, sure enough, there was a figure in overalls walking quietly away from the door. It was obvious that he was a painter who had been painting windows from a cradle on the outside wall, and when it started to rain he had simply opened one of the fourth-floor treatment room windows, walked within 2 feet of the terminals of the 400 kV tube and gone about his business. He never knew how close he had been to a very unpleasant accident. I am not sure what would have been the reaction of a modern radiation protection adviser.

3.2 X-ray and Other Equipment

We might pause here in the meandering around medical physics and look briefly at the X-ray and other equipment which was available to us in the 1930s. It is not at all easy to conjure up now the appearance of a radiological department of 60 years ago. The old gas X-ray tube, deriving directly from Röntgen's original apparatus, had largely disappeared, though a small number remained in use. I can think of one which continued for use in animal experiments until 1940. These had been replaced for radiodiagnosis and therapy by the Coolidge tube with its bright glowing filament and solid tungsten anode, usually mounted in a lead-lined box, with an appropriate aperture. Even around the time of which we are thinking, the open Coolidge tube was on its way out to be replaced by the revolutionary Metallix tube, the forerunner of almost

Mercury breaks (A E Dean and Co.), reproduced from Mayneord 1929 The Physics of X-Ray Therapy (J & A Churchill).

all modern X-ray tubes for medical use. This was the brainchild of the great Philips X-ray engineer, Albert Bouwers. Instead of a complete but fragile glass tube, there was a strong metal cylinder forming the middle part of the X-ray tube, welded at each end to a heavy glass cylinder, into which the anode and cathode were sealed. The central metal cylinder could be earthed and carry a beam-defining metal shield and other equipment. It could also be mounted on a fixed or mobile metal frame. All this now seems so obvious and crude, but it signalled the advent of the modern radiological department layout. By this time, the specialist engineers had taken over from the DIY enthusiasts who had dominated the first quarter of the century.

In the field of generating apparatus, the induction coil, which had been the main source of high voltages for radiology, had almost disappeared. A high-voltage induction coil could be a very frightening piece of equipment, which leads to a very personal comment. I must have some claim to fame as one of the very few surviving people who can remember having the duty of 'gassing the interrupter'. The interrupter was a substantial steel cylinder inside which a rotating mercury jet made intermittent contact with two electrodes wired into the primary circuit of the induction coil. To reduce arcing and burning, an inert atmosphere inside the cylinder was desirable, but as nitrogen and carbon dioxide were rather expensive it was the practice to keep the cylinder filled with town coal gas, kept under slight positive pressure from the mains. This was fine so long as no oxygen filtered in. The real danger arose if air leaked in overnight when the pressure was turned off. The possible consequences of this were horrendous and there was a strict rule that, before switching on each morning, the interrupter must be fully gassed, ensuring that it was full of gas and not a lethal gas–air mixture. This was normally the duty of a designated radiographer or occasionally the junior physicist. A very good description of this monstrosity appears on page 157 of Mayneord's little classic book, *The Physics of X-Ray Therapy*.

The 10 years or so leading up to the start of the Second World War saw a complete transformation in the appearance of an X-ray department. To those of us on the ground much of the change seemed to happen without our realising what was going on. Exposed X-ray tubes, electrically live wires and bars quietly disappeared and the latter were replaced by insulated cables, even up to 200 kV. Fully screened and insulated X-ray tubes appeared, whilst transformers and rectifying valves disappeared into the bowels of huge oil-filled tanks. Within a very few years, X-ray departments for both diagnosis and therapy became much safer places in which to work. Radiological equipment had rapidly developed the 'black box' image. In laboratories far removed from hospitals, studies of the physics of the solid state were producing dramatic results. Solid state rectifiers became widely used in small-scale, radio-type equipment and it was not long before they were upgraded for use in high-voltage radiological equipment. The old brightly shining valves were gone, leaving the X-ray tube as the only remaining thermionic component.

There was a further development in therapeutic equipment in the 1930s which had a profound effect on treatment practice, particularly during the war. This was the Metropolitan Vickers 250 kV therapy machine. In

Metropolitan Vickers 250 kV machine items: continuously evacuated X-ray tubes (top) and control room (bottom).

this, the fragile glass X-ray tube was replaced by a heavy steel cylinder, some 35 cm in diameter, with an earthed tungsten anode sealed in one end and a large porcelain insulator, carrying the filament gear, at the other. The system was vacuum sealed and continuously evacuated by pumps mounted on the same mobile frame as the tube.

At the time this looked like an almost impossible innovation but it worked remarkably well, having at least two considerable advantages. If anything went wrong, the whole system could be dismantled, repaired, reassembled and re-evacuated. The other advantage was that, unlike the conventional fragile X-ray apparatus, it was virtually unbreakable. As an example of this, one night at the Middlesex Hospital, a German bomb exploded not more than 30 metres from the Metropolitan Vickers equipment and caused great local damage. Next morning, the therapy rooms were cleaned, the X-ray machines examined and treatments were resumed by mid-day. Indeed, though now long gone, these most unusual machines played a most important part in maintaining a radiotherapy service in several parts of Britain throughout the war. After 50 years, I can now confess that I tried to go one better than Metropolitan Vickers. I designed and built a small, continuously evacuated X-ray generator on Metropolitan Vickers lines in which both X-ray tube and rectifying valves, not semiconductors, were in the same evacuated chamber. For a short time it actually worked, but the bomb mentioned above demolished both laboratory and equipment and ended the investigation.

3.3 Radium

In radiotherapy, there was, of course, one other source of radiation for treatments, namely the gamma rays from radium-226. The larger radiotherapy departments usually had about 1 gram (1 curie, 37 gigabequerels) of radium, sealed in hollow platinum needles and tubes, with contents between 1 and 20 mg. Some of these collections are, I believe, still in existence. These needles and tubes were used to cover the whole range of so-called brachytherapy, interstitial, intracavitary and surface moulds. In London, smaller departments had access to a substantial pool of needles and tubes, maintained by The King Edward's Hospital Fund under the care of Professor Russ. The Radium Pool staff also provided an informal advisory service on the physical and dosimetric aspects of radium therapy. An alternative to radium, particularly for interstitial work, was radon gas. At several sites around the country, a gram or more of radium chloride was held in carefully shielded solutions; from this, radon gas was milked off each day into fine glass capillary tubes, which were then cut off into 1 cm lengths. These were then enclosed in thin platinum sheaths to become 'radon seeds'. With the short half-life of radon, seed implants could be, and often were, left *in situ* within the body permanently. Again London was favoured. The King's Fund maintained a centre at the Middlesex Hospital Medical School from which radon could be ordered and delivered within the region. Looking round at the senior hospital physicists today, retired and retiring, it is quite surprising to realise how many were introduced to the field via the Radium Pool or the Radon Centres.

3.4 Hunting for Radium

Almost any hospital physicist from, say, pre-1950 will have one or more horror stories arising from the clinical use of radium and particularly of radium searches. Of necessity, needles and tubes had to be carried around frequently between radium safe, sterilising room, operating theatre, wards and laboratories. Although most careful checks were demanded, opportunities for loss were not uncommon. Radium tended to disappear down drains, into surgical dressings and hence to waste bins, on then to incinerators and garbage tips. Any such disappearance amounted to an emergency, as, apart from the monetary value of the radium, it represented a degree of public hazard. Most physicists of the period could tell tales of searches in laundries, waste bins, smelly sewers, incinerators and corporation garbage tips. I could tell of half a dozen, starting with the Huddersfield adventure recorded in Chapter 2. To digress for a moment, I have a clear recollection of a day of doubled horrors, one of which has survived for more than 50 years. It was wartime and one Sunday morning I was called out to search for lost radium in a North London corporation dump and incinerator. After several very dirty hours, the search was partly successful and I was invited to lunch at the hospital concerned (I suspect The Royal Free!). Food rationing at that time was very severe, but lunch smelled delicious—steak pie apparently. Disillusion came very heavily, when I realised that it was whale steak, my very first experience of that dubious delicacy of wartime meals.

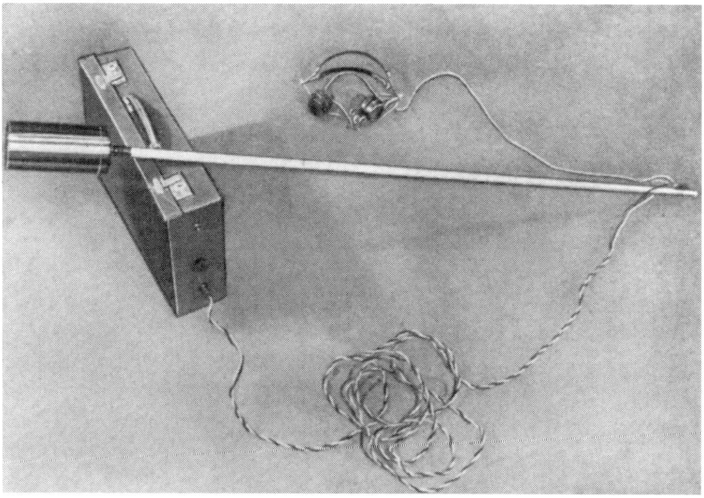

'Clucking Hen' radium detector, reproduced from Wilson 1948 Radium Therapy (Chapman and Hall).

In the earliest days, the only instrument sensitive enough to detect a few millicuries of radium at a reasonable distance was a gold leaf electroscope. I think it was Mayneord who considerably increased the sensitivity of the standard laboratory electroscope by attaching a metal plate, 10 or 15 cm in diameter, to the pillar holding the gold leaf, rather in the form of an ion-collecting aerial. The sight of a wretched hospital physicist crawling around a laundry drain or a garbage dump, charging up the electroscope and waiting for the leaf to fall, was indeed depressing. Relief came when Tom Chalmers invented the 'Clucking Hen' of blessed memory. Although not quite a true comparison, we thought of this as a poor man's Geiger counter. It consisted of a small neon discharge lamp wired in a resistance–capacitance circuit, with a battery of a voltage just below the normal firing voltage of the lamp. When the gas in the lamp was ionised by a particle, the firing voltage was reduced; there was a discharge and a flash. Put a pair of headphones in the circuit and in the presence of ionising radiations we have persistent clucking. Looking for lost radium was never the same again and life was distinctly easier, but radium searches never became the most popular exercise for the hospital physicists of the time.

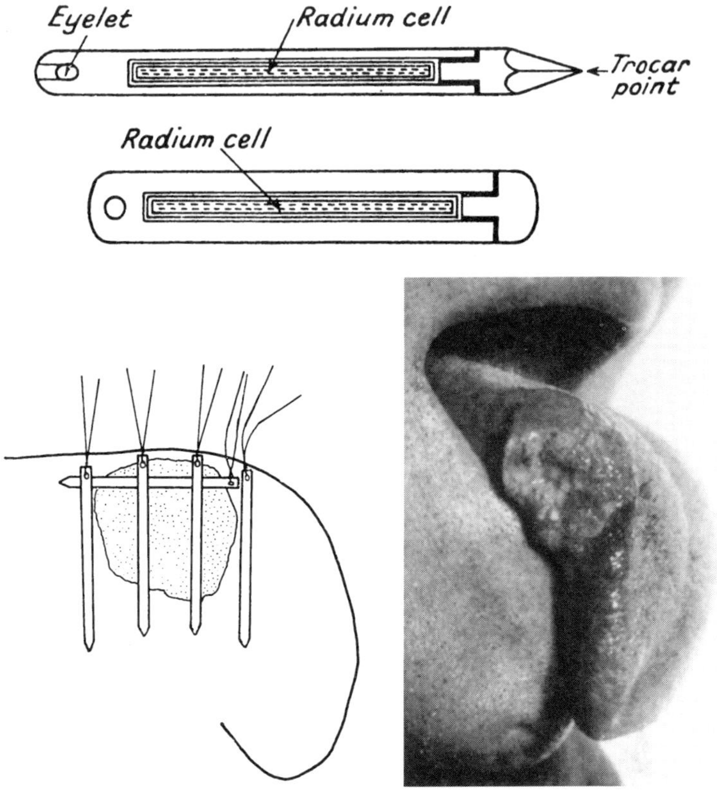

Early radium needles and tubes and an example of their use in treatment of a tongue cancer, courtesy of Dr R F Mould.

3.5 Radium 'Bombs'

Nowadays, when we think of telecurie units, we naturally think in terms of very large machines containing kilocuries of, say, cobalt-60. True, many of these have now been replaced by very high-voltage accelerators. It is interesting to note, however, that they are at the end of a long line. Telecurie units, known earlier, rather unfortunately, as radium bombs or later cobalt bombs, have been in use for well over 60 years. There was one such 'bomb' at The Cancer Hospital as early as

1930, and I shudder to think of it. It consisted essentially of a cube of lead of about 15 cm side, mounted on something very like a portable X-ray stand. Into one side of the block was cut a 5 cm square aperture towards the centre of the block. The whole of the Hospital's available stock of radium needles and tubes, around 1 or 2 curies, were wired onto 5 cm square plastic cards and these were packed in the bottom of the 5 cm square hole and fixed there. Thus from the 'bomb' there emerged a 5 cm square section gamma-ray beam and when the unit was in contact with the skin, the approximate source–skin distance was 5 cm. This was surely the parent of all succeeding telecurie units. It is difficult to imagine a cruder arrangement but it must be said that the old radium bomb gave excellent service, particularly in the treatment of difficult head and neck cases. There was, however, one very serious snag: the whole of the hospital's radium stock could not be devoted to this work for the whole of the time. Every day, needles and tubes were required for other forms of treatment. Thus, every morning all radium which had been removed from patients' treatments was assembled in the radium room, together with all requests for radium to be used in treatments that day. The 'bomb' was duly opened and returned tubes and needles rewired to the appropriate cards and the day's requested sources taken out. Everything then had to be counted and checked before the unit could be sealed and returned to clinical use as a 'bomb'. There was no elaborate remote handling or control equipment. A simple, 2 inch thick lead table, L-shaped, had to suffice. Radium was handled with long-handled forceps or, for wiring onto cards, with tweezers. Several people were authorised to carry out this procedure but, somehow, it fell with great frequency to the junior physicist. Fortunately or not, we had no such thing as a radiation protection adviser. One curious consequence of this procedure was that the dose rate from the 'bomb' varied erratically from day to day. After the morning 'opening', each patient's dose chart had to be adjusted by the daily correction factor.

At this point, I usually tell the story of how the radium bomb almost wrecked my own marriage before it even started. During the night before I was due to travel to Yorkshire for the wedding, there was an accident and the whole unit was overturned. I was called in the early morning and spent most of the day, when I should have been travelling north, putting things to rights. I made it to Yorkshire eventually, even managed to get married—just.

A small number of a new generation of telecurie bombs came into use from about 1935 onwards, designed originally by Dr L Grimmett of the Radium Institute and built by Messrs Bryant Symons. Radium sources of 5 or 10 grams (curies), sealed in 5 cm diameter light-alloy cylinders, became available. Such a container was stored in a heavy lead-lined safe in the treatment room. The treatment head would be recognisable as a smaller version of a modern cobalt unit. The treatment head and the safe were joined by a flexible metal tube, some 2 or 3 metres long, through which the radium source could pass. The patient was set up for treatment whilst the radium was still in the safe, and after the operator had retired the source was blown through the tube to the treatment position and the exposure started. At the end of the pre-set treatment time, the radium container was sucked back into the safe. The equipment was wonderfully simple: an ordinary domestic vacuum cleaner produced the blow and suck via a simple two-way valve. It worked remarkably well. This elegant apparatus never became widely used, probably because of limited radium supplies and, of course, soon after the end of the war, cobalt-60 solved most of the problems.

3.6 Development, Maintenance and Quality Assurance

The development, maintenance and what we would nowadays call 'quality assurance' of the equipment and techniques, in those early days of radiology and radiotherapy, placed a heavy burden on the small number of physicists employed in medicine. Yet, as is essential in any scientific discipline, time had to be found for more fundamental research and development. There were a number of relatively fundamental problems on the agenda, mainly concerned with the interaction between high-energy X-rays and gamma rays with matter, absorption, scattering and energy distribution. This all now seems very unsophisticated until one realises the then state of knowledge of high-energy radiation interactions. It is difficult for the present-day radiological physicist to realise that in the mid 1920s there was no obvious explanation of how the absorption of hard gamma rays in tissues could produce any significant biological effect. Yet even then, these rays were being used to treat cancer. Apart from the photoelectric effect, which at these energies is negligible in tissues, the only other known form of 'absorption' was Rayleigh scattering, which left no energy in the scattering medium. The Compton effect, whereby energy is deposited

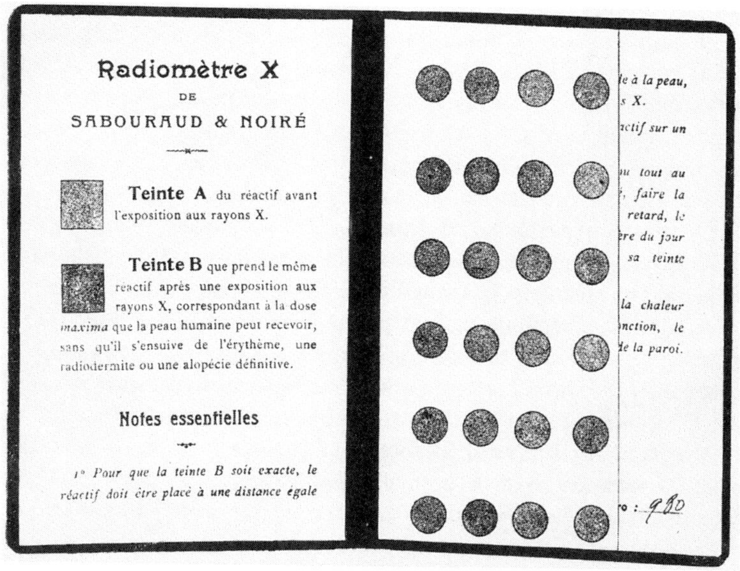

Sabouraud–Noire pastilles, courtesy of Dr R F Mould.

during the scattering process, was only discovered in 1923 and was one of the most up-to-date and exciting topics of my own undergraduate physics course.

3.7 The Röntgen: the Unit of 'X-ray Dose'

We should realise that, in spite of the difficulties which plagued the physicists who wished to measure X-rays and set up a system of dosimetry, radiotherapy was already a flourishing speciality. Dosimetry there was but it was somewhat primitive and arbitrary. There was a mysterious quantity known as the 'mean erythema dose', defined roughly as 'the amount of radiation which will just produce a mild erythema on the average patient'. I must leave to the imagination the difficulties and problems in applying such a 'quantity' in everyday physical and clinical practice. In practice, each radiologist had to derive from experience the necessary exposure from the radiologist's own X-ray tube, which would produce this average erythema. The calibration of the X-ray tubes was effectively carried out on the patient's skin, subjected to that skin erythema dose. It may nowadays seem quite indefensible but it must be said that thousands of patients were treated, many very successfully, on this basis, surely a tribute to the clinical judgement of many radiologists. There was one physically based alternative which had not quite disappeared when I entered the field, namely the Sabouraud–Noire pastille dosemeter. This came in several forms, but the Sabouraud–Noire was most commonly used in Britain. The pastille was a small plastic disc about 8 mm in diameter, covered with crystalline barium platinocyanide. These crystals were bright green in colour but took on a brown coloration when exposed to X-rays. The depth of the coloration was measured against a series of pieces of tinted glass. The tints were expressed in terms of fractions of 1 B, which was supposed to correspond to one erythema dose. The pastilles returned to their normal colour when exposed to daylight. Experts claimed that these doses could be read to within 10 per cent or better but I have to confess that I could rarely achieve a repeatability of better than 30 per cent. The system had many obvious disadvantages, apart from any question of accuracy. It was, of course, necessary to interrupt the treatment from time to time, remove the pastille and read the dose. Pastilles varied in sensitivity with temperature, humidity, dose rate and so on. More serious, perhaps, was the fact that it was necessary to place the pastille half-way between the X-ray tube target and the patient's skin. The horrors of an overdose arising from an innocent or ignorant operator placing the pastille on the skin are very obvious. In spite of all this and much more, the system was used with some success for some years. I can claim that in 1939, long after I had left The Cancer Hospital, I was asked to calibrate one X-ray machine in pastilles. This was because one very distinguished dermatologist, who shall be nameless, insisted on prescribing doses for skin malignancies in these units. My accuracy had not improved beyond the 30 per cent range. There are other reports of pastilles being used as late as 1945.

During the 1920s there was a growing realisation that ionisation, probably in air, was the most promising phenomenon for the measurement of X-rays, whatever may be meant by measurement in this context. The literature shows that quite a number of laboratories, not all hospital based, were working on the subject, devising quite a range of ionisation measuring systems and suggesting quite a number of possible units of quantity or dose. There were the German R-unit, Solomon's R-unit (not quite the same), Friedrich's e-unit, Duane's N-unit and several others.

By 1928 there was a considerable degree of agreement and at the 2nd International Congress of Radiology in that year, the röntgen was agreed and defined as the unit of 'X-ray dose'. At this point, the quotes around the word dose are very necessary as, writing in the 1990s, the uncertainties which have arisen over the years about the definition of such terms are only too obvious. Much of the credit for the acceptance of the röntgen was due to the interest of the national laboratories, the National Physical Laboratory at Teddington, the National Bureau of Standards in Washington, DC, and the Physikalische und Technische Reichsanstalt in Berlin, all of which promptly set up standardising equipment. The unit was confirmed at the 3rd International Congress in 1931 and by 1932 was widely accepted in radiotherapy departments throughout Europe and the United States. As that was the year I entered the field, I might very self-consciously claim that radiological physics had at last come into its own.

3.8 Measurement Problems

There were, however, still fundamental problems, some of which persist today. Some commercial dosemeters were available, especially the beloved Hammer and Mecapion integrating instruments, but a good deal of improvisation in dosimetry was the order of the day. Rolf Sievert must have been working on his famous or infamous condenser ionisation chambers—small, hollow light-metal spheres, about the size of a good marrowfat pea, with a tiny insulated spherical electrode inside. Loss of charge from this electrode, measured with a sensitive electrometer, was a measure of dose to the chamber. These typified the ionisation problems we faced before first thermionic electronics and then solid state electronics took over.

A major fundamental problem was looming. Calibrations in röntgens were limited to a maximum photon energy of 200 kV, and were reasonably satisfactory over the range 50–200 kV, covering most of the immediate needs of radiotherapy. There was, however, a growing demand amongst therapists and physicists for higher energies. They had little doubt that higher energies and deeper penetration must lead to improved clinical results. As noted earlier, The Cancer Hospital had acquired two sets to work at 400 kV; there were whispers of millions of volts in the United States and at Bart's in London. Mayneord had built a free air ionisation chamber much larger than usual, which promised to serve the purpose of measurement in röntgens up to 400 kV, but probably no further.

3.9 Dosimetry Research at The Cancer Hospital (Free)

There was, of course, one other radiation being used in radiotherapy, namely the gamma rays from radium (B + C), and these spread well into the million electron volt range. A more satisfactory system of measurement of these radiations in radium therapy was an urgent requirement. Up to this time, although radium and radon were widely used, dosimetry was virtually non-existent or, to be charitable, somewhat arbitrary and subjective. The only available physical unit was the exposure of the radium in milligram-hours. To specify a treatment in full, there had to be added details of spatial distribution of the radium or radon tubes and the distances from skin or other organs. These details could only be empirical, deriving entirely from the skill and clinical experience of the radiotherapist. What was required, of course, was some measure or estimate of the dose, preferably in röntgens, at points in the treated body.

To the hospital physicist of today, this situation must seem incredible. The measurement of gamma rays in röntgens loomed large in the programmes of almost everyone with pretensions in medical physics, including our little group at The Cancer Hospital. The thing we aimed at was the dose rate in röntgens per hour at 1 cm from a point source of radium with, say, a 0.5 mm platinum filter. From earlier ionisation measurements, this could be calculated and most calculations suggested a figure between 8.0 and 8.5 röntgens per milligram-hour. However, when the measurements were made with available free air ionisation chambers, figures around 2.5 röntgens per milligram-hour were obtained, a serious discrepancy to say the least. It was soon realised that the cause of the discrepancy lay in the range of the secondary electrons which produced the ionisation, roughly what every trainee hospital physicist today knows perfectly well. For us, at The Cancer Hospital, the answer came via a series of very simple experiments. Apparatus was very basic. Some good electrometers, particularly the Lindemann, had become available, but for the moment we preferred our simple gold leaf electroscope, in which electrometer and ionisation chamber were incorporated into one. Our basic instrument had a 5 cm cubic celluloid frame, covered by graphited cellophane, enclosing the usual insulated rod. There was a gold leaf hinge of about 112 mm by 1.5 mm onto which we glued a graphited glass fibre about 1 cm long. I write 'we' deliberately; there was friendly competition between Mayneord and myself for who could produce the best leaf–fibre system. The fibre was observed in a simple telemicroscope. Having measured the capacitance of the electroscope and produced a voltage calibration, it was relatively simple to estimate the ion production from the rate of fall of the leaf. I have deliberately introduced this touch of basic elementary physics to illustrate the gulf between our approach to the problem of dose measurement 60 years ago and that of modern sophisticated varieties of dosimetry.

It soon became clear that the key to the problem lay in the wall thickness of the electroscope, which was, of course, the ion chamber. Using a carefully collimated beam of gamma rays, the ionisation increased with wall thickness of celluloid to a maximum at about 5 mm. On the other hand, the same experiment, without collimation of the beam, produced a falling curve reaching the same steady value at the equivalent of 5 mm celluloid. The key was that, to measure the true ionisation in the cavity, it was necessary to enclose the air volume with an equilibrium thickness of air-equivalent material, equal to the maximum range of the secondary electrons. Described 60 years later, it all sounds rather like sixth-form physics, but for us it was the key to a desperately needed system of dosimetry. The crucial point was that, assuming air equivalence of the celluloid, the measured ionisation in the chamber led to the most important figure of 8.3 röntgens per milligram-hour at 1 cm. We were not alone in this move forward. The National Physical Laboratory (NPL) approached the subject from quite a different direction. NPL built a free air ionisation chamber about the size of a modest lecture room and eventually came to the same conclusion and the same key figure. As mentioned in a later chapter, I was allowed to present these results at the International Congress of Radiology in Zurich, 1934, and must confess this was probably the most exciting experience of my scientific life.

3.10 Dose Distribution

In addition to the major dosimetric problems, we managed to find time for other, perhaps rather less urgent, researches, most of them concerned with the more precise applications of radiation treatments of patients. We were very concerned, for instance, with the secondary radiations around radium applicators. When Sievert produced his remarkable small ionisation chambers, there was a great burst of activity working out the distribution of dose around a variety of radium arrangements. I remember, too, studying the spectra of high-voltage (250 kV) X-rays with varying filtration, using the old Zeeman spectrometer of blessed memory. If any such instruments are still in existence, they must be museum pieces with almost antique status. I might end this chapter with an account of one small piece of research work, which, whilst not world shaking, was very exciting.

3.11 Absorption Coefficient Measurement

Having access to what were then high-energy X-rays, we were interested in precise measurements of absorption coefficients. By heavy filtration of a 250 kV beam, we were able to produce almost monochromatic radiation of around 59 XU (equivalent to 210 keV, 0.059 Å). We then measured absorption in about 20 elements. This confirmed what we already knew or assumed: that for light elements up to sulphur, where photoelectric absorption was negligible, the mass absorption coefficient was the same at 0.120 $cm^2\ g^{-1}$, for all elements, giving an absorption per electron of around 4.0×10^{-25}. Hydrogen is, of course, the exception, having double the electron density of all the other elements, but we could not demonstrate this directly. The hydrogen content of water should, however, increase the Compton mass absorption by some 11 per cent and, sure enough, we could demonstrate this. Then, one day in 1935 a mysterious parcel was delivered to the Physics Department. It contained a phial of some 75 ml of water, but what water? It was, in fact, deuterium oxide, heavy water, a thing of great wonder to us, straight from Norway. We had no conception of heavy water in great bulk as moderator for a nuclear reactor or for anything else; we had no real conception of a nuclear reactor—this was 1935. The phial was carefully stored in the safe until we were ready to measure its X-ray absorbing powers. When we did this, the result was not unexpected but nevertheless exciting. The mass absorption coefficient for heavy water was back at 0.120, as for all other light elements except hydrogen. We had proved that deuterium had a 'normal' electron density. I must apologise for using the 1935 nomenclature in these matters; it seems right that way. This effort rated a letter to *Nature* and nowadays I have to smile at our rather pompous conclusion: 'It may be observed that absorption measurements of this type serve to count electrons per atom, as distinct from the more usual method of weighing isotopic nuclei', surely the enthusiasm of scientific youth.

3.12 Conclusions

Looking back on those first few years of my own life in medical physics, I must admit it looks a little like the Dark Ages. Nevertheless, our problems were very real and urgent; it was a new experience for physical scientists to be spurred on by clinical urgency. Incidentally, as I read my *Physics in Medicine and Biology* and my *British Journal of Radiology*, every now and then I see questions propounded which are almost identical to some put to us 60 years ago; often I have thought we had answered them. Some of the now proffered answers do not seem very different.

CHAPTER 4

NEW HORIZONS

From a conventional historical point of view, the period from the mid 1930s to the mid 1940s was a time of unrest, disturbance, rumours of conflict and eventually worldwide war itself. This was not exactly the state of affairs in which one would expect a new and peaceful scientific discipline to take off. In spite of it all, however, medical physics and its associated technologies showed, during that time, modest, if not spectacular, growth and development. The seeds sown in London's Fulham Road, in Manchester, Stockholm, New York, and a few other places, were beginning to show their first fruits. At that time and for some time to come, medical physics was still dominated by the needs of radiology and particularly radiotherapy. Until the war intervened, there was steady and sometimes dramatic development in the techniques of X-ray production and use and this was reflected to some extent in clinical developments in both diagnostic and therapeutic radiology. For the time being, we few physicists were firmly attached to radiology. Without daring to stray too far from my own field, it is perhaps reasonable to suggest that cancer therapy was largely in the hands of the surgeons; they even owned personally a good deal of the available radium. At the same time, radiotherapy was beginning to make a substantial impact, not so dramatic as in post-war years but still impressive. Not very far ahead were the days when surgeon and radiotherapist would come together in joint clinics and just occasionally might invite a physicist in for consultations.

4.1 The Middlesex Hospital

At this point in my travels, I should explain how my own circumstances and pilgrimage were changing and developing, At the end of 1936 I meandered from Chelsea to London's West End proper and this became

The Middlesex Hospital in 1935.

the centre of my activities for the remainder of my career. In that year the Middlesex Hospital was planning to open a new department of radiotherapy, one of the largest in the country. Radiotherapy was to be separated from the Diagnostic Radiology Department and, in fact, the two departments were at opposite ends of the main building. To take charge of radiotherapy was a young and enthusiastic Brian Windeyer, lately out of Australia and fresh from a period working with the radiotherapy wizards of Paris. With this great expansion, he insisted he needed a full-time physicist. After some heart searching and negotiation, I was offered and accepted the post. I can now admit that it was with considerable hesitation that I left the burgeoning research-based excitement of the specialist Cancer Hospital for the unknown world of a large general teaching hospital but, with long hindsight, there is no doubt that, at least for me, it was the right move. Some people will realise that, being the Middlesex, the story was not quite so simple. In the matter of laboratory-based services, the Middlesex had always worked to an almost unique system. The Hospital itself has never had laboratories for pathology, biochemistry and so on. These services were provided by the Medical School on a contra-account basis, the heads of departments becoming members of the hospital staff. Thus, after he completed his Beit Fellowship in 1913, Sidney Russ, the pioneer of us all, was appointed Physicist to the Hospital; later, when he moved over to the Medical School to take over the newly created Joel Chair of Physics, he retained his place on the staff of the Hospital, and, as such, appeared in the Medical Register.

All this was made very clear to me in writing on my appointment. I was quite definitely appointed to the academic staff and discipline of the Medical School Physics Department and would be seconded to the Hospital Radiotherapy Department to give advice and assistance. It is only right, however, to point out here one great advantage of this strict attachment to the University department. Within reasonable limits, it was possible to devote any available spare time to academic research, not necessarily aimed at the direct requirements of clinical medicine. At that particular time, this was very much an academic matter, as spare time was in very short supply, but I came to realise the importance of this freedom in later years.

The new Radiotherapy Department which was being established at the Middlesex was typical of what was going on in many places throughout the country in those immediately pre-war years. By the standards of the day, it was a very large and modern department, with its own entrance foyer and occupying a whole wing of the older hospital. By that time, fully insulated X-ray generators and tubes up to 200 kV and 10 mA were available and four conventional Newton & Wright sets were installed on one floor. On another floor, there were four of the new Metropolitan Vickers continuously evacuated steel and porcelain tubes with generators up to 250 kV. As pointed out earlier, these were a godsend in the darker days of the war when bombing was a real hazard. They were virtually indestructible. On the ground floor was installed one of the Bryant Symons telecurie units (I am afraid we called them bombs), containing all of 10 grams of radium. Also on the ground floor was a smaller therapy set working between 50 and 100 kV for the treatment of more superficial conditions. Those were the days when, every now and then, the physicist would be asked to plan a multiple field treatment which would entirely depilate the head of some wretched child, so that ringworm could be treated. Today, this sounds horrendous but somehow we usually managed to carry out the treatment without long-term disasters. To complete the armoury, another gram or more of radium in needles and tubes was available for brachytherapy (a word which had not then arrived). The whole made up a very powerful therapeutic force. With two large, rather old-style wards, some smaller rooms, space for outpatients, reception and records, it constituted a considerable departmental complex.

It will be noted that in this description of a major radiotherapy department, there is no mention of physics. Being the Middlesex, this was elsewhere, but not far away. The Barnato Joel Laboratories, which housed most of the Medical School Physics Department, adjoined Radiotherapy and I was allocated a large and convenient laboratory for my section of the work. I must admit that some people thought this allocation unfortunate as it deprived their cancer patients of their billiard room and table. On that point I had to be neutral. In addition to the laboratory, I was allocated the then large sum of £100 for any additional instruments or equipment I might need. To that sentence I must add that the Physics Department had a very well equipped and staffed workshop, though with little or no electronics.

For the time being, I was nominally the only full-time physicist allocated to this large department of radiotherapy, although the situation was not quite so bad as that sounds. Apart from the professor, there were several members of the Physics Department staff with sufficient experience to bail me out if needed. There was also the group of young physicists, mentioned earlier, who staffed the King's Fund radium and radon services. A quick glance at the list of founder members of the HPA throws up quite a number of names of people who served in these processes, names such as Jack Boag, Alan Jennings, Sidney Osborn, Clifford Walker, Frank Stewart and others.

4.2 Treatment Planning

During this period of the late 1930s and the 1940s, radiological physics became much more patient orientated than in the earliest years. There was a story passing round the Middlesex and elsewhere, which may well be apocryphal, that no physicist there should ever be allowed to see a patient. Details of a patient's condition and the required treatment should be set out in writing and sent to the Physics Department. In due course, written details of the proposed treatment plan would be sent back to Radiotherapy. Perhaps a pinch of salt should be prescribed for this item. As I settled into the Department, the newly commissioned equipment meant increasing demands for calibration, maintenance, quality assurance—all the now familiar routine chores of a radiological physicist. It would be difficult to define the point at which the routine planning of individual patient treatments began. Clearly it was introduced gradually and, for X-rays, was only possible when isodose curves for most of the available treatment regimes were forthcoming. This was, in itself, no mean task, as the HPA depth dose tables and the Meredith–Neary isodose calculations were still some way into the future. Curves had to be produced individually on the machines in out-of-hours working time. Gradually, however, the physicist became a partner in the radiotherapy team, attending clinics to collect information and data and to make measurements necessary in the planning of a treatment. In the medical climate of the period, this was almost revolutionary.

By present-day standards, our planning of individual treatments must seem crude in the extreme. Patients' outlines were obtained using a simple piece of lead 'rope' and transferred to an 18 inch by 24 inch sheet of paper, together with relevant anatomical information, including the radiotherapist's estimate of the shape and position of tumour tissues. It was then the business of the physicist, using isodose curves and a large measure of imagination, to try and optimise the multiple beam treatment of the sketched-in tumour volume. Everything was, of course, in two dimensions; computers and computer programs simply did not exist. It was, indeed, very crude but we had our successes and being informed of clinical successes and failures, in itself a most unusual experience for the layperson, was a most satisfying aspect of the business. Even at this distance in time, I have a clear recollection of a period when one of our ENT surgeons was collaborating with Radiotherapy in treating a series of cases of neoplasm in the facial sinuses. We, the physicists, were given as much anatomical and other information as could be established from radiographic and clinical examinations and from this we had to plan multiple field irradiations with curative intent. We were working very much in the dark. Weeks later, the surgeon in question would usually carry out exploratory surgery and tell us in the clearest terms whether our treatment plans had been a success or a miserable failure. It was a most salutary experience. Electronic storage of information and data was a long way in the future and retention of treatment plans was only possible in the physical form of the large sheets of paper on which they were drawn. We thus accumulated enormous albums of these plans covering most areas of the anatomy and a whole range of treatment conditions. These became a valuable asset in the process of optimising succeeding plans and we became quite expert in declaiming the 'I seem to remember a case like that' syndrome. How we would have envied the full computerisation of treatment planning and the electronic storage of plans and data—or would we?

The problems of planning treatments with radium or radon did not seem quite so severe. By the late 1930s the use of the röntgen as the unit for prescribing gamma-ray dose was well established and for practical applications, we had the Paterson–Parker tables. These were a brilliant concept based on a very simple idea and were of immense value to generations of struggling medical physicists. Surface moulds with radium or radon were widely used; the therapist would prescribe dose and timing, leaving the physicist to determine the sources to be used and their distribution. The idea of a dedicated mould room for radiotherapy had not yet appeared but the physicists and radiographers did their best with moulds. Interstitial implants involved a similar system. From data provided by surgeon or radiotherapist, the physicist produced a suggested planned distribution of the needles to be inserted. In our system, this plan appeared as a diagram on the wall of the operating theatre. The rule was that the physicist must be in the theatre to interpret the plan and make any on-the-spot changes demanded by surgery or anatomy. My main objection at that time was that radium operations were always scheduled for 8.30 am and as I lived an hour's journey away, such activities were never popular. The use of radium in gynaecology, mainly in the treatment of carcinoma of the uterine cervix, caused little trouble. British radiologists were divided between two systems of radium distribution and dosage: those of Paris and Stockholm. Not surprisingly, the Middlesex was Paris orientated. I seem to remember being asked to determine from radiographic and other data the dose at 'Point A', but I must admit that today I would be hesitant to define exactly where 'Point A' was to be found.

Paterson–Parker tables, reproduced from Meredith (ed) 1947 Radium Dosage (E & S Livingstone).

Volume (cm^3)	Milligram-hours per 1000*r*	Volume (cm^3)	Milligram-hours per 1000*r*
1	34.1	110	783
2	54.1	120	830
3	70.9	140	920
4	85.9	160	1005
5	99.7	180	1087
10	158.3	200	1166
15	207	220	1243
20	251	240	1317
25	292	260	1390
30	329	280	1460
40	399	300	1529
50	463	320	1595
60	523	340	1662
70	579	360	1726
80	633	380	1788
90	685	400	1851
100	735		

Elongation factor	Elongation correction
$1\frac{1}{2}$	+3%
2	6%
$2\frac{1}{2}$	10%
3	15%

Useful filtration corrections	
Filter used	Correction to radium used
0.3 mm Pt	+4%
0.6 mm Pt	−2%
0.7 mm Pt	−4%
0.8 mm Pt	−6%

4.3 Expansion at the Middlesex

The size of our new Radiotherapy Department and the increasing sophistication of the physics services demanded soon made it clear that the load was too much for one physicist, even with help from the Medical School. An increase in the medical physics staff was inevitable, an increasingly common phenomenon around the country. I have mentioned earlier the development and installation of the rather special radiotherapy machines by Metropolitan Vickers. For this work, they had recruited a team of young physicist–engineers under the care of Dr T E Allibone. Several of these young men became hospital physicists, the best known being George Innes, who brought the first million volt machine to Bart's. One of these engineers, Raymond Quick, joined me at the Middlesex. He was with us for some years and then went to St Georges, London, and Addenbrookes, Cambridge, before crossing the Atlantic to become Chief Physicist and later Assistant Professor of Radiology at Strong Memorial Hospital, Rochester, NY. A little later, just after the outbreak of war, Frank Farmer was tempted from his prestigious post at Marconi to come to the Middlesex. This was a most important catch for the growing profession of medical physics, as many who know the physics departments of the north-east of England are well aware.

4.4 Farmer-type Dosemeters

This item inclines me to muse about the way in which one small flash of inspiration can transform an everyday problem. At the time in question, we were concerned with many routine problems of X-ray dosimetry, which nowadays are rightly considered long ago solved. In particular we were dealing with the provision of built-in permanent output meters, with an ionisation chamber in the head of a radiotherapy machine. We could make ionisation chambers of any shape or size; we had electrometer-type thermionic valves; but for constant everyday use, there were serious problems of stability. Anyone working in the field of electronics of that time would know all about feedback in valve amplifiers. When, in a moment of inspiration, Frank Farmer suggested we apply a large degree of negative feedback to the electrometer circuit, this led to the design of a stable direct-reading instrument for measuring ionisation currents using a small thimble chamber on the end of a long flexible cable. Later, Farmer developed a dosemeter based on the Townsend balance principle which transformed the stability and convenience of X-ray dosemeters for the next 50 years or more. It is difficult to overestimate the part played by Farmer-type dosemeters in the standardisation of routine X-ray dosimetry in all parts of the world (see sections 13.2 and 13.7).

CHAPTER 5

A GROWING BUSINESS

In the previous chapters we have been largely concerned with my own activities and meanderings through the period 1932 to 1945, when medical physics began to grow and expand. This personal reminiscence should now be placed against the larger background and the wider fields of activity. The burgeoning growth of the use of ionising radiations in the treatment of malignant disease was clearly the main factor in the increasing demand for the services of physicists in clinical activities. This was clear in the developments at Fulham Road and at the Middlesex and these were widely reflected in other places throughout the country and in some other countries. These rather ponderous thoughts on the development of medical physics and radiotherapy lead to a rather curious observation, which may come as a surprise to some modern practitioners. When we now think of the growth of radiation therapy, it is usual to visualise X-ray machines of ever increasing size and power, with radium or, nowadays, other radioactive elements playing a supplementary role. Nevertheless, the upsurge in radiotherapy and hence of medical physics in this period was not primarily concerned with X-rays but rather with the development of the therapeutic use of radium.

5.1 Radium Regulation in the UK

We could probably date the beginning of the expansion—we might even equate it to the end of the DIY period—as around 1930. Prior to this date, much of the radium available for clinical use was in the hands of individual surgical teams, some of it even owned by the surgeons; control, safety, dosimetry were, to say the least, a little arbitrary. X-ray therapy, where it was available, was in the care of radiologists, most of whom had commitments in diagnostic radiology. Radiotherapists, as such, were very thin on the ground; the qualifying diploma (DMR) of the time covered both specialities. A point of take-off came in 1929. The government of the day set up the Radium Commission, provided cash for the purchase of radium and empowered the Commission to promote radium therapy, controlling the use and safety of that dangerous element. Between 1929 and 1936, the Commission set up about two dozen National Radium Centres, mainly in provincial centres of population. London was omitted in the earlier stages as radium was available to many hospitals from the King's Fund, in the care of Professor Russ. There were also some five Medical Research Council Centres with available radium. One of the major conditions under which radium was provided was that in each centre there should be a 'fully trained' therapist as 'Radium Officer'. As mentioned above, the only formal training was in the combined Radiology Diploma. I might quote from Gordon Spears' fascinating history of the Commission: 'the Commission pressed for adequate recognition and remuneration for the officer in charge of Radiotherapy, and his status during this period developed from that of "House officer" to membership of the consultant staff'. I might add a reminiscent comment to this. When I joined The Cancer Hospital in 1932, the official 'Radium Officer' there was the unforgettable Constance Wood and she had reached the elevated rank of 'Radium Registrar'. Such were the trials and struggles of the newly developed speciality of radiotherapy and this was closely linked to the development of our profession of medical physics.

From the point of view of these meanderings in medical physics, the crucial years were probably 1936 and 1937. In its 1936 Report, the Radium Commission noted with apparent approval that a number of the National Radium Centres had appointed physicists to their staff. In the following year, the Commission went further. It decided that a physicist should be added to the list of personnel essential for the working of a National or Regional Radium Centre. It was pointed out that it was important to appoint physicists trained in the applications of radiation in therapy and in the associated calculations. This is the first reference I can find to the training of hospital physicists and suggests a comment. In the succeeding years, hospital physicists played a leading role in the development of training for 'Radium Officers', eventually radiotherapists, but 40 years elapsed before we had a comprehensive scheme for the training of physicists themselves in their aspects of radiotherapy and other medical disciplines. This point in the activities

Founder Members of the HPA: Louis 'Hal' Harold Gray, reproduced from History of the Hospital Physicists' Association 1943–1983.

of the Radium Commission is well illustrated by taking a look at the list of Founder Members of the HPA.

5.2 Physicists in Medicine

In 1932, physicists directly attached to clinical departments in Britain numbered perhaps five or six but by 1934, things began to move. In that year, Hal Gray moved from the Cavendish Laboratory to Mount Vernon to start a career which had a profound effect on the scientific and professional standing of medical physics. About the same time, Charles Wilson moved from the Strangeways Laboratory in Cambridge to Westminster Hospital where he was joined for a time by John Read. It was there at Westminster that he joined that extraordinary background character in medical physics, Harry T Flint, the academic physicist-cum-radiologist mentioned in Chapter 1. With his dual career and

Founder Members of the HPA: Henry 'Harry' Thomas Flint, reproduced from History of the Hospital Physicists' Association 1943–1983.

contacts, Harry Flint was indeed a most important influence. In those advances in the 1930s, Manchester and the Christie Hospital could not be far behind, and around that time appointed Herbert Parker as physicist. I clearly remember our first contacts. Herbert was sent to The Cancer Hospital to see what radiotherapy physics was all about and then to go on to Stockholm to see Sievert at the Karolinska Hospital. One of Herbert's troubles was that his travel and expenses allowance was very limited. As a Yorkshireman, I should not comment on such a Lancastrian problem. He spent much time in London trying to find a cargo ship which would take him to Sweden at minimum cost. In the following year, after a period of moonlighting efforts at Lambeth by Mayneord and myself, the London County Council appointed L H S Clarke from the Middlesex to take over the heavy task of organising radiological physics at Lambeth and Hammersmith Hospitals. He was quickly joined by Denis Jones from the National Physical Laboratory. In 1936 also, the best known of Allibone's Metropolitan Vickers team,

George Innes, joined the staff of Bart's and started work on Britain's first million volt X-ray generator. Meanwhile, Scotland could not be left out and another of the Metropolitan Vickers team, Charles Murison, was appointed to The Royal Infirmary in Edinburgh.

This, then, was the scene which so impressed the Radium Commission and determined that a physicist was a must in the organisation of a Radium Centre. The trickle of appointments continued until wartime conditions slowed it down, only to become a torrent after the war. We could see the beginnings of groups which were to become the major departments of medical physics in the post-war world. When I moved to the Middlesex, my replacement at The Cancer Hospital was James Clarkson, who had been with Phillips in Birmingham. He was soon joined by Len Lamerton. In 1937, the National Physical Laboratory produced another medical physicist and sent Harold Cook to Bradford. Jack Meredith joined the Christie Hospital and eventually succeeded Herbert Parker, who had been tempted away to a distinguished career in the United States. At Mount Vernon, Gerry Neary joined Hal Gray, though he may be better known in radiological physics for his collaboration with Meredith in the calculation of isodose curves; he later gained greater fame in radiobiology. Tom Chalmers left the academic precincts of Bart's to set up a new department of medical physics in the Liverpool Clinic; in London, Walter Langmead set out his stall at St Thomas's. The list may not be quite complete but it may be enough to illustrate the point. Although I have steadily denied any intention of trying to write a history of medical physics I have to admit that the previous few paragraphs look suspiciously like a parade of honour of the real pioneers. History or not, it has been an honour and a privilege to meander in such company.

A major factor in the employment of physicists in the hospitals was the change in the organisation of radiotherapy. The Radium Commission was set up to deal with the clinical uses of radium, leaving X-ray therapy to the radiologists. By the mid 1930s, the weakness of this arrangement was apparent, and the Commission began to encourage the uniting of the disciplines and the setting up of specific departments of radiotherapy. Surprising as it may seem, this process was only slowly implemented; there were clearly professional problems both here and abroad and the change was only completed here in Britain by the time the Radium Commission was wound up and the National Health Service took over in 1948. By that time, the growth of medical physics had accelerated and was spreading into other parts of the medical field.

5.3 Marking Time

Inevitably, the outbreak of war brought disruption to the hospital services as well as to personal lives. During the war years, a small number of new names were recorded in the medical physics services but, on the whole, it must be regarded as a period of marking time. Whenever possible, radiotherapy services continued to operate normally, though some patients were moved from major urban sites. Fortunately, the developments in the production of new X-ray apparatus for therapy had peaked at an opportune time and much of the apparatus in use was relatively new. This was particularly the case with the new Metropolitan Vickers continuously evacuated X-ray generators. These proved a saving factor in many places.

Radiotherapy and its associated physics at the Middlesex carried on at a somewhat reduced rate owing to reductions in staff. Many members of the technical staff were called up for war service and research and development came almost to a standstill. At one stage, in desperation, I set up a very small watchmaker-type lathe in my own room and, more or less self-taught, became relatively expert at producing miniature ionisation chambers. Teaching and the production of new doctors had to go on and the pre-medical section of the School was moved to Bristol at the start of the war and then in 1940 to Leeds. At that time I was drafted to Leeds to take charge of physics teaching and so left mainstream medical physics for a short time. Although upsetting at the time, this proved to be a most interesting and, in view of later developments, a very valuable experience. I was even allocated my old room in the University, the room I had left some 10 years earlier. For a short time, I was able to leave behind the harassment of everyday clinical problems and enjoy the atmosphere of academic research with old colleagues. However, after a year, it was thought I would be more useful in London and I was quickly brought down to earth. In the later stages of the war, life indeed became very hectic. With depleted staff, days at the Hospital were long and these were often followed by long nights on Civil Defence duty; I even reached the exalted rank of corporal in the ARP.

5.4 Radium, Sponges and Syrup

As pointed out, we managed to keep radiotherapy running reasonably well although there were few technical advances to report. Radium, however, presented many problems. It was realised, long before war broke out, that there could be a serious risk from the dispersal of radium by a high-explosive bomb or shell. A number of measures were taken to reduce this risk. Wherever possible patients were moved to hospitals in lower-risk areas away from large centres of population. In most places, protection of the radium stock had to be provided. A cylindrical steel 'bomb proof' container some 18 inches long and 12 inches in diameter, with walls 3 or 4 inches thick, was designed and constructed. This could easily hold a gram or more of radium in needles and tubes, with little external radiation. I understand these containers were very thoroughly tested in secret against high explosives and proved very effective. For larger quantities of radium, such as we had at the Middlesex, a somewhat more ambitious method was used. Encouraged by the Radium Commission, some hospitals were provided with a borehole some 50 feet deep and about 18 inches in diameter. A cylindrical brass container, somewhat narrower than the hole, with several compartments, was used to carry the radium down, using a simple hand-turned hoist. The general idea was that all radium treatments should be confined to daylight hours and that so far as practicable, radium should be underground at nights or during air raid alerts. We had one such borehole at the Middlesex, housed in a small building near the basement entrances of Radiotherapy and Physics. It was well used, though it produced some anxieties and occasionally a little amusement. Some non-Middlesex radium was housed in our borehole together with the hospital's entire stock of needles and tubes and also the 10 grams from the telecurie unit described earlier. When any radium had to be collected or deposited, the whole stock had to be brought up, a significant hazard, as we could not hope to keep the ambient dose rate below accepted safety levels. We did our best. The 10 gram telecurie source was a real headache. Each morning at about 8.30 am, this source had to be brought up, placed in a lead pot and carried by hand on the run to the treatment room to be inserted into the safe. It was similarly returned each evening and if there were an air raid alert, the whole procedure had to be repeated. The very idea of carrying 10 curies of a powerful gamma emitter by hand through hospital corridors would, I know, horrify most present-day safety officers, but, at the time, there was no simple alternative. For most of the time, I was responsible for these operations, but there was never time to make the desirable measurements of hand and whole-body dose. So far as we know, no one came to any significant harm. There were lighter moments. Details are vague, but I seem to remember it was reported that someone had dropped a small quantity of non-Middlesex radium down the borehole without the assistance of the cylinder. How does one recover a couple of small needles from a hole in the ground, 50 feet deep and 18 inches in diameter? My recollection is that the most promising method involved a series of canes of the chimney sweep or drain rod type, with a large sponge on the end, coated with golden syrup. Whether any radium was ever recovered I never knew. My suspicion is that in several places in the country, there are significant radium sources lying 50 feet underground. The good news of this story is that no radium was lost nor dispersed as a result of enemy action. Unfortunately, the headquarters of the Radium Commission was totally demolished by a bomb in 1940.

5.5 Conclusions

This brief and rather personal outline of the activities in medical physics in wartime cannot end without reference to what is probably the most important event in the history of the profession. In 1943, Britain and the greater part of the world were in the throes of the greatest war in history. Britain itself had suffered months of intense aerial bombardment of military and civilian targets, the country was tightly blockaded from the sea, food and many other things were rationed and in very short supply. Hospital staffs were depleted and most hospitals were running on an emergency basis. This description of wartime conditions could go on. The fact is that this was the time chosen to form the HPA, to represent the profession to the world. A full story of the launch and progress of the Association is told in a volume published by the HPA for the 40th anniversary in 1983, and it will not be repeated here. Almost inevitably, the initiative came from Professor Russ, who, as we have seen, started it all in 1909, together with the other appropriate professors in London, Hopwood, Stead and Flint. I myself should have a niggling complaint about this business. Although I was at that time one of the longest-serving hospital physicists, I knew nothing of it until some time later. Then it was disclosed that the original invitation was limited to heads of departments, with a proposal for a forum in which such seniors could discuss matters of common interest. Assistant physicists were out and I was very much the assistant. A strange recollection comes to mind here. Apparently at the first meeting, Charles Wilson and I were nominated for the post of first secretary of the HPA. Charles was quite rightly appointed. I first heard of this nomination episode 50 years later, at the time of the Bristol Golden Jubilee Congress in 1993, when I saw a copy of the 1943 Minutes in the *HPA Journal*! In the original proposal also, professional and political matters were to be excluded; we might have had a very different HPA. In the event, all these limitations were firmly removed, largely by the efforts of those energetic young men who had been appointed to hospitals in the north of the country. In fact the Northern Branch of the HPA came to life many months before the 'parent' body, if such an obstetric miracle can be contemplated. The HPA and its later more academic progeny, the Institute of Physical Sciences in Medicine, have played a most important role in my subsequent meanderings in and out of medical physics. This should appear from time to time in these notes. The latest metamorphosis, the Institute of Physics and Engineering in Medicine, which still flatters me with an Honorary Fellowship, came too late significantly to affect my wanderings. It has clearly become a body of considerable power and influence. For the moment, I must just repeat the quotation I used in the 40th anniversary booklet: 'They builded better than they knew'.

CHAPTER 6

POST-WAR EXPLOSION

As may have been gathered from previous chapters, in the eight or nine years up to 1945, with its unrest and war, medical physics was, to some extent, marking time. This was quite inevitable under wartime conditions. Numerically, the profession grew very slowly, passing the 50 mark in Britain before the war ended. During this time a number of people came in who were to play a most important part in the development of the profession. Indeed, as we have seen, right in the midst of war, the HPA was formed. Although some of us may have had our doubts, this proved to be a very fortunate piece of timing. It meant that there was an organised body in position to react to the changes which were to come in the field of physics applied to medicine. During the years immediately following the war's end, the expansion of the physical sciences into medicine was quite explosive and just a little chaotic. In the 5 to 10 years from 1945, the whole situation was changed; our immediate attachment to radiotherapy remained firm but avenues of service and advance appeared over almost all the medical spectrum. I would suggest that only now, in the 1980s and 1990s, can we see anything like the full extent of the explosion. This is perhaps best seen in the Institute of Physical Sciences in Medicine (IPSM), born out of the HPA as a learned society to advance its scientific activities, both academic and clinical. Indeed, even as some of these meanderings were being recorded, the explosion went on. The IPSM has now long gone to be replaced by the Institute of Physics and Engineering in Medicine, the outcome of a real coming together of the IPSM and the Biological Engineering Society. The range of the activities of the new Institute in publishing, scientific meetings and conferences, and linkage with other scientific and medical bodies, can only be described as amazing. Dreaming back to the relative simplicities of the life of a member of the HPA around, say, 1945, we can only respond, 'Whatever have we done?'

6.1 Professor Sidney Russ Retires

As I am supposed to be recording my own meanderings, perhaps I should start this phase with my own experiences. Soon after the end of the war, my own personal and professional position changed quite dramatically. After more than 35 years at the Middlesex, Sidney Russ, the unquestionable pioneer of modern medical physics, was due to retire from the Joel Chair in 1945. In view of the disturbed war situation, the University in fact suspended the rules and postponed the retirement to 1946. This was clearly the end of an era, just when a period of explosive development was almost inevitable. In the event, the University took no steps to fill the vacancy until the beginning of the 1946–1947 session, and I was installed as temporary head of department.

That period from October to December 1946 is one I would prefer to forget. Naturally I had ambitions to make the temporary appointment permanent but there could be no certainty about this; my fears were all the greater when, legally or otherwise, I saw a copy of the shortlist for interviews. Perhaps fortunately for me, the candidate I most feared withdrew, having found a Chair elsewhere in London. The University eventually installed me in the Joel Chair of Physics Applied to Medicine, and I fancy I was walking on air.

It is very difficult more than 50 years later to appreciate what a daunting prospect lay before me, despite my immediate pleasure. Perhaps the most significant fact of all was that I had to follow Sidney Russ.

Russ was a most complex character and I would not dream of attempting an analysis. There can be no doubt about his tremendous influence on the early development of medical physics as an independent discipline. I was always impressed by his ability to operate in all aspects of the machinery of medicine and particularly at the higher levels of hospital hierarchy. Much as I respected Sidney Russ, we did not always see eye to eye, neither in scientific or technical matters nor in professional affairs. There was one occasion when he told me quite openly and without ceremony that I would never succeed him at the Middlesex. He was wrong, but only just, and my respect for his life and influence and for his memory remain.

6.2 J E Roberts: Joel Professor of Physics Applied to Medicine

Willy-nilly, then, at the end of 1946 I was plunged into a most exciting and demanding situation. During the war years, the old Physics Department had been ticking over, maintaining its traditional functions of teaching, hospital services and just a little research work, with a minimum of staff and resources. The time had come for a major overhaul and, almost inevitably, a massive expansion, applied in all aspects of the Department's activities.

The first duty of a professor is presumably to teach. The classes for the First Medical (1st MB) and, to a lesser extent, the Radiological Diplomas had been steadily maintained. In the first two or three sessions following the end of hostilities, there were quite dramatic changes in the make-up of the classes, particularly for 1st MB. The schoolboys of 18 or so (girls came a little later) were outnumbered by young men of more mature years, whose education had been interrupted by the war. Their outlook had not been developed in the classrooms and playing fields but rather on battlefields, on the high seas or in the air. Most of them were desperate to get on with life and qualifications and they were prepared to work hard for them. There was real pleasure in teaching such classes. In one matter they were defeated. At the end of one session they expressed their thanks by inviting me to lunch, which really meant a stand-up snack at the bar of the local pub. The immediate appearance of two very large tankards on the bar in front of me implied clearly that the real hope of the class was to see the new professor of physics legless. It was indeed a disappointment when they discovered that the said professor was effectively a non-drinker. Perhaps I was a little disappointed myself.

During the war years, the physics service to the Hospital and particularly to radiotherapy was remarkably well maintained and in some ways advanced. Radiation dosimetry was definitely improved, largely due to the genius of Frank Farmer. In radiotherapy, treatment planning remained a desktop, pencil and paper business but experience was accumulating which was invaluable when computerised planning was introduced. Much of this was made possible by a most interesting list of young physicists who passed through the Department between, say, 1939 and 1955. I hope to mention these in detail later. A simple calculation shows that, including Professor Russ and myself, the Middlesex provided fully 20 per cent of the founding membership of the HPA.

Inevitably during wartime, research was very limited. Teaching and hospital services occupied most working hours and essential outside services filled any spare time. Professor Russ's radiobiological research work continued but was in the process of being run down. My own tentative efforts at an unusual X-ray generator, which perished under an enemy bomb, have already been mentioned. Because it could be carried on beside hospital work, I managed a rather academic study, linking the gamma-ray spectrum of radium (B + C) with experimentally observed absorption data. The time was ripe for a major overhaul of the Department's programmes.

6.3 Building a New Department: Physics Applied to Medicine

In 1946, the field of medical physics was wide open both in the advancement of clinical services and in a wide range of physics research programmes with a clear medical flavour. Wartime research for defence purposes was about to pay dividends in peacetime life and services. We were challenged effectively to make a fresh start. So far as the Middlesex was concerned, my first requirement was to find an experienced senior colleague to help face this challenge and to build up a new department. In this I was extremely fortunate. Harold F Cook, one of the most underrated hospital physicists of our generation, had just been released by the Royal Navy. True, his immediate experience in that august company had been as a gunnery radar officer and he could occasionally be persuaded to tell stories of controlling the guns of a cruiser, particularly when bombarding the French coast on D-day. Some 13 years earlier he had started his professional life at the National Physical Laboratory under G W C Kaye. As early as 1937, with little or no hospital background, he moved to The Bradford Radium Institute, to set up from scratch a successful new physics department. Harold came to us in 1947 and for the next 27 years was a tower of strength for me; his wise counsel saved me time and again from doubtful or even false steps. He came to a department which, for some time, took on a distinctly naval flavour. Several of the new scientific and technical

Harold Frederick Cook, reproduced from History of the Hospital Physicists' Association 1943–1983.

staff had served in the Royal Navy and both the departmental secretaries were ex-Wrens. There was a tendency to refer to my study as the old man's cabin and to HFC as Number One.

We spent many hours trying to peer into the future for a glimpse of our new Department of Physics Applied to Medicine. The undergraduate teaching programme was in sore need of revamping; we even had the audacity to insert some applied medical physics into the first- and second-year pre-medical courses. The demand for physics services from clinical departments, particularly radiotherapy, continued to increase.

In a way, it was fortunate that the development and wider distribution of megavolt accelerators for X-rays and of large telecurie units was just a few years ahead. This gave time for the development of dosimetry and particularly for the expansion and improvement of techniques for the individual planning of X-ray and gamma-ray treatments. Most exciting was the opportunity to set up an entirely new research programme. The results of the enormous research and development programmes arising from the war effort were increasingly being made available. We were not alone in the scientific world in looking at these results, with dreams of applying them in our own fields.

Two of these areas of research seemed to us ripe for further development and application in the field of medical physics. The most obvious to us and to many others were the developments arising from the Manhattan Project, with its nuclear bombs. Almost before the war was over, experimental nuclear reactors were producing small amounts of radioactive nuclides and some of these were being made available for medical investigations. The effect on our Middlesex department was almost overwhelming and it would seem a good idea to allocate what we came to call nuclear medicine to a later chapter. The other field which interested us was that of microwave radiation in the 1 to 10 cm wavelength range, which had been produced at high power levels for use in radar. This was an entirely new section of the electromagnetic spectrum which would surely have some clinical applications and probable uses in biological research.

This sounds very vague and naïve for the planning of a major project but I think it fairly represents our position at the time. As a first step, we purchased from RAF surplus stores a complete H2S radar transmitter just as it came out of an aircraft. This purchase prompts a small diversion; I am reminded of a small item of meandering which we made several times during the late 1940s. Following the war, the British government had enormous quantities of technical, mainly electrical and electronic, equipment surplus to requirements. This was stored, piled up high in huge aircraft hangars near Oxford. The equipment was made available to schools, colleges and universities at give-away prices. We made several visits to the stores and our own store cupboards became full to bursting.

6.4 Microwave Generators

Returning to our microwaves, we soon realised that for some purposes we would need a much more powerful source of radiation. After negotiations with the Royal Navy (I suspect Harold Cook had something to do with all this) there appeared in the yard outside my room a very large trailer van. This contained a large high-powered microwave generator, used by the Navy for radar gunnery control. Before long, waveguides were erected, the microwaves were piped into the laboratory and we were in business.

I suppose our ideas about possible direct clinical applications must be regarded as very simple. We thought that at least physiotherapists would be interested in microwave heating. High-frequency heating was commonly used and the extension to microwaves seemed obvious, particularly as they could be directed and localised. Our first effort then was to produce a defined beam of microwaves from the end of a waveguide, very similar to a radiotherapy applicator, and then develop a system of dosimetry or perhaps more strictly thermometry in tissues from the absorbed beam. We were clearly greatly influenced by our experience in radiotherapy. A conical applicator was fixed on the end of a waveguide, and we had a very familiar set-up. Just for curiosity, we boiled (coddled?) eggs in containers placed in the beam; were we the first microwave cooks? What we needed were isothermal curves produced when a body was placed in the microwave beam. An inanimate, tissue-like phantom, as used for radiotherapy, was clearly not suitable; some account had to be taken of blood circulation. It was clear that live phantoms, preferably human, were the answer, a considerable mass of living tissues such as a human buttock being the obvious solution. The 'dosemeters' for this exercise were hypodermic needles of various length (depth), incorporating thermocouples. As might be expected, the response to our appeal for volunteers to produce our tissue isothermals was not a very enthusiastic one. There was an order extant (from whence?) that the Head of Department should not suffer this indignity. Despite the difficulties, reasonable temperature distributions were obtained but, unfortunately, the clinicians never took up the technique. Nevertheless, we learned a great deal about tissue heating and the associated hazards of microwaves. This was of considerable interest to the radar fraternity and other users of microwaves. Harold Cook, who supervised all this work, became a much consulted authority in the field of microwave hazards. It is of interest to reflect that many years later there was a strong demand from radiotherapists for a system of tissue heating to use directly or in combination with radiotherapy in the treatment of some malignant conditions. It may seem that we missed the boat.

6.5 Other Activities

The advantages of being part of an academic institution as well as a service department became obvious in this particular field of microwaves. The dielectric properties of biological materials had never been a field of great clinical interest but they offer fascinating scientific problems. Edward Grant came to us from King's College, London, to study for a PhD and, with the equipment we then had available, soon found biological dielectrics, particularly those of hydrated compounds, a most fruitful field of research, publishing quite a range of papers. This work continued when he moved to Queen Elizabeth College and later back to King's College, London, taking a world lead in this fascinating field of biophysics. In another direction, we were rather naïvely interested in the possibilities of some form of molecular resonances in this newly available spectral frequency range. Harold Cook and Tom Buchanan set up a programme to investigate such resonance frequencies, with only limited success. Maybe we missed the boat again. We may, with hindsight, feel that it was fortunate that Tom Buchanan took his expertise to join John Mallard in Aberdeen to continue this type of work. We know that, some time later, there emerged nuclear magnetic resonance imaging followed shortly by spectroscopy, developments from medical physics which are clearly changing the face of diagnostic medicine.

While all this research work and also that in the field of radioactive isotopes was going on, there were increasing demands from radiotherapy and other hospital departments for physics services. Radiation protection became a subject of increasing importance and the greater sophistication of treatment planning, though it would nowadays be considered rather primitive, placed a heavy burden on the physics staff. Most, if not all, staff members had to share their available time between clinical demands, research work and some teaching. Between, say, 1946 and 1960, we built up what, in my vanity, I would regard as a most impressive team of young men and women. The word team reminds me that, to my great delight, for several years, we did indeed have enough young men, scientists and technicians in the Department to field a very promising football (Association) team. On several occasions we played the medical students' team, the Biochemistry Department and others. Being in my forties, I was considered much too old to play, but I was occasionally allowed to referee, much to the disgust of the biochemists.

The Old Professor and the New: J E Roberts (on the right) with J F Tait on his appointment to the Joel Chair.

6.6 A Training Ground for Medical Physicists

The reminiscences of that period are almost certainly the most satisfying of my life. In this long-distance remembrance there is an impression of a small number of departments which had become nurseries for the supply of hospital physicists for the increasing demands at home and abroad. My Middlesex team (physics, not football) was just one part of this movement or, as I have called it in the chapter title, this explosion. Reminiscent pride is such that I must, even at the risk of some embarrassment, fall for the geriatric temptation to list some of the members of that team who contributed so much to medical physics. Harold Cook did not go out from the Middlesex; he resisted the temptation of a university chair elsewhere and stayed to build and maintain our Department. Quite a number found university chairs in many places. Mention has already been made of Edward Grant, who ended up as Professor of Physics and Head of Department at King's College, London. James F Tait, FRS, came to the Department having followed exactly the same track as myself from the University of Leeds.

Founder Members of the HPA: Roy Ellis, reproduced from History of the Hospital Physicists' Association 1943–1983.

Although he was involved in the whole range of departmental duties, his main claim to fame was the discovery and characterisation of the adrenal steroid aldosterone, as recorded in Chapter 7. James eventually moved to the United States where he acquired a worldwide reputation for his researches in the field of endocrinology, but in 1970 returned to the Middlesex to succeed me in the Joel Chair and as Physicist to the Hospital. James was, I believe, the first medical physicist to become a Fellow of The Royal Society and he and his wife were the first married couple ever to be so designated for the same piece of work. Much better known to hospital physicists was Roy Ellis who was with us from 1954 to 1964, though he took a year off to go to New York as Secretary of The United Nations Committee on the Effects of Atomic Radiations. Later he became Reader at Bart's and eventually succeeded Bill Spiers at Leeds. Edward Williams was mentioned as early as page 1. He came to us as a possible PhD student but switched to medicine and ended up as the Director of The Institute of Nuclear Medicine and also Professor. Amongst my list of professorial successes, I find the story of Roger Ekins most intriguing. He came to us soon after the war on the strength of a very modest Natural Sciences Tripos and the ability to drive a car for the King's Fund Radium Pool. He stayed on and worked as a hospital physicist but his genius blossomed in nuclear medicine, where he became Reader in Medical Nucleonics and eventually Professor of Biophysics. His methods of microbiochemical analysis are very widely known. Another name already mentioned and well known in medical physics, with his name on many dosemeters, is Frank Farmer. After some years at the Middlesex he went north to found a great medical physics dynasty in Newcastle upon Tyne and became a professor in the University of Newcastle.

Others of our physics team moved out to head the departments of medical physics which were growing up in the post-war years. I go back to Sidney Osborn, who was in the wartime King's Fund group and who moved first to University College Hospital and then to be Director of Medical Physics at King's College Hospital, London. Norman Ramsey went out to head the Physics Department at Charing Cross Hospital and Douglas Mabbs to a long career at Reading. Further afield, I still hear from and have visited Kenneth Clarke in Australia, now retired as Chief Physicist at the Melbourne General Hospital. Although he did not head a hospital department, Clifford Gregory went on to become for a time Chief Scientist at The Ministry of Health, a post recommended by the Zuckerman Committee. Two members of the team left hospital physics as such for the world of commerce. Paul Cook established a business which eventually became one of the largest manufacturers of lasers in Europe. Ralph Worsnop (yes, his father was, with Harry Flint, the author of *Practical Physics*, beloved by generations of physics undergraduates) emigrated to Philadelphia and later to California and set up a profitable agency producing computer software for radiological treatment plans.

From time to time there were overseas members of the team; two come immediately to mind. Engeline Gouw-Lee spent some years in the Department before returning to Singapore to lecture on radiation physics at the University there. Her presence in Singapore proved most useful in my own travels in that part of the world. Kemal (Kim) Mahmoud was one of the outstanding characters of the team (physics and football).

He came to us from the Egyptian Atomic Energy Authority and, after obtaining a PhD, returned to that Authority and eventually to a Chair in Cairo University. He was, I think, largely responsible for my own affection for that city. There were quite a number of others from the Middle East, Hong Kong and elsewhere.

Looked at in this way, it seems not a bad score for a more or less concurrent group in a modest department: seven university chairs, four or more heads of major hospital physics departments, one Chief Scientist and two successful business men. To all this must be added one Fellow of The Royal Society and two quite separate Nobel Prize nominations for different pieces of work. There may be others I should have mentioned but memory becomes fragile.

6.7 Conclusions

I have written of those who joined our team at the Middlesex and then moved on to spread the message of medical physics more widely. Of course, some came and stayed the course to retirement. I have already written of Harold Cook. I think also of Tom Bryant, the mainstay of our radiotherapy physics for many years; there was John Tappin, who started in audiology physics and went on to wider fields; there was John Bradley, a great teacher and research worker who, nevertheless, remains in my memory as the surprising author of a Chinese–English scientific dictionary. There were others but most of them a little too late in my own career to comment upon. I trust the people mentioned will forgive my intrusion into their records.

The Middlesex was not alone in this developmental explosion. Around the country other major centres were bringing along the young men and women who exploded that small, rather odd group of so-called hospital physicists of the 1930s into a major profession with its own association and learned institute. Similar movements occurred in other parts of the world, though maybe not quite so dramatically, so that now we have national, continental and worldwide organisations for medical physics. The visions which Mayneord and I saw as we sat in the night hours by our 400 kV X-ray machine in 1933 have come to fruition beyond our wildest dreams. It has been a privilege to have been a part of this long and fascinating birth and growth of a new profession.

CHAPTER 7

NUCLEAR MEDICINE

This rather bare chapter heading should not be misunderstood. It is now some 50 years since radioactive isotopes became fairly readily available and the time may well be ripe for a definitive history of nuclear medicine. This, however, is neither the time nor the place for such an effort. Suffice it at this stage for these meanderings in nuclear medicine to be effectively restricted to our little patch at the Middlesex.

7.1 The Development of Radioactive Nuclides

Even before the war there were, of course, tentative experiments with man-made radioactive nuclides but these were rare and radioactive materials in very short supply. Only when information about the atomic bomb was released and experimental nuclear reactors came on stream could we begin to dream of the future of nuclear medicine. Most of the established hospital physics departments quickly jumped on the isotope bandwagon and our Department at the Middlesex was no exception. There was tremendous enthusiasm and by 1948 we were offering a limited isotope service to the clinicians, mainly with iodine-131, phosphorus-32 and sodium-24. By today's standards the techniques available would seem primitive in the extreme; for example, our almost universal radiation detector was a very basic end-window Geiger counter. Fortunately the peculiar physiology of the thyroid gland and the ready availability of radioactive iodine-131 were godsends to the first practitioners of nuclear medicine. They became the foundations of a whole range of clinical investigations. Better and better determinations of the uptake and distribution of iodine activity in the thyroid were an obvious objective for most isotope departments. I have clear recollections of our first efforts to produce a thyroid scanner.

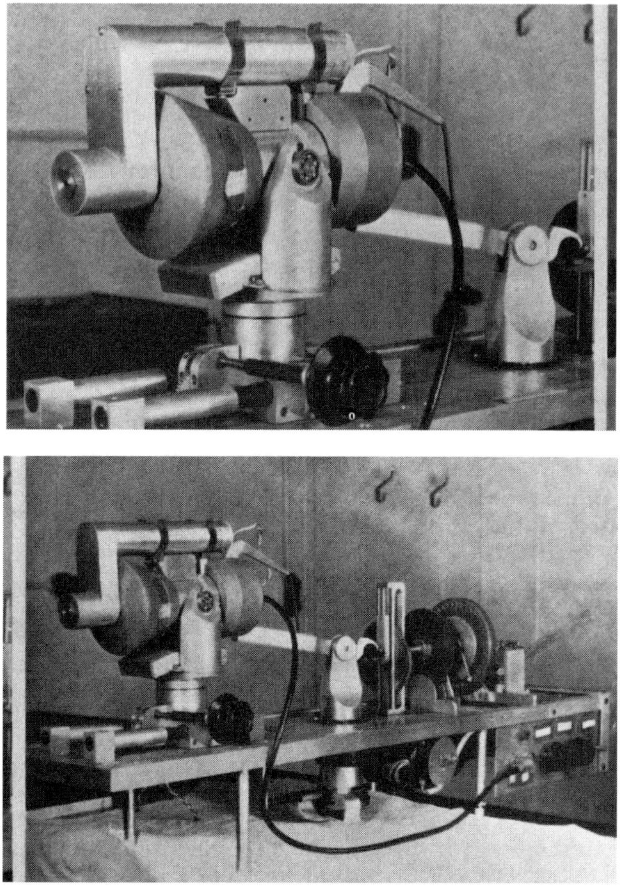

'Universal Radium Detector': Geiger counter and optical devices, reproduced from Mayneord and Newbery 1952 'An automatic method of studying the distribution of activity in a source of ionizing radiation' British Journal of Radiology, Volume 25, No 299, p 590.

A single end-window Geiger counter with a lead collimator could be moved around a rectangular frame by a system of hand screws. This was fixed in front of the neck of a more or less immobilised patient who had been given an appropriate dose of iodine-131. Moving the counter over the rectangular area, counts of about 1 minute were taken at intervals of 5–10 millimetres. Thus, after an hour or so of counting, a kind of 'isodose' curve of iodine activity in the thyroid could be produced. At a slightly later time the system was motorised and counting automated; we had a scanner, though by no means a pioneer. From such crude beginnings, the modern department of nuclear medicine has its gamma cameras and their increasingly sophisticated offspring for imaging. I have a very clear recollection of one of the very first hand scans of a thyroid which we carried out with our prehistoric machine. At that time, some surgeons were distinctly reluctant to subject their patients to this 'potentially dangerous' radiation technique. In our case, the Professor of Surgery had decided to do a thyroidectomy on the basis of his own diagnosis and agreed we could do no harm with our scan on the previous day. We carried out the laborious exercise and found a large inactive hole in one of the lobes of the gland. We reported this as a possible cyst and awaited the consequences. The dissected thyroid showed a cyst exactly where we had predicted and from that day the reputation of the Department of Physics rocketed, at least in Surgery.

7.2 The Isotope Unit

From 1950 onwards the isotope work increased steadily both in quantity and the range of investigations undertaken. However, the range of tests we could offer to the increasing demands from the Hospital was still rather limited, although we were able to maintain a fairly extensive research programme. At the same time the demands for physics services from the Department of Radiotherapy and other clinical departments increased steadily; there was also an increasing consciousness of the hazards of radiation, and all these factors together were steadily increasing the service load on the Department. Increases in staff became inevitable and names like Tait, Worsnop, Ekins, Williams, Tappin and others appeared on the staff board. We even had to accommodate chemists Whitehead and Reith to deal with the intricacies of isotope chemistry. I must record how fortunate we were during this period of expansion that Brian Windeyer was Dean of the Medical School and that he recognised the significance of these developments. Neither the Medical School nor the Hospital resisted the growth of this strange new isotope unit.

As well as the increases in staff, the advent of unsealed radioactive materials required a replanning of the laboratory arrangements. An urgent need was a high-activity laboratory for the dispensing of higher-activity isotopes for clinical use. By present-day standards our methods and equipment were again very primitive, but we were very conscious of the hazards these radioactive materials presented. Thinking of our little high-activity laboratory brings to mind a strange irrelevant anecdote. One day when our dispensing laboratory was very new, I was proudly showing it to a small group of scientists from the Soviet Union. As I was talking, Sylvia Simpson (later Tait of aldosterone fame), who was working with us at the time, came in and I introduced her. I shall never forget the look of amazement on the faces of our visitors when Mrs S immediately started up conversation in fluent Russian, a rare achievement at that time. Radioisotope work involved a good deal of chemistry and several well-established physics laboratories required a face-lift for this purpose. More difficult still, a considerable portion of the clinical work involved *in vivo* investigations, requiring the presence of the patient. Few physics laboratories would be considered suitable for this kind of work and much refurbishing and redecoration was required.

In this matter of accommodation for nuclear medicine, the Middlesex was particularly fortunate. The Physics Department, in which the Isotope Unit was nurtured, had been housed from its inception in the Barnato Joel Laboratories, attached to the main hospital buildings. In 1959, just when the accommodation problem became really acute, the new Windeyer Building of the Medical School was opened to house the pre-clinical departments of the School. Physics moved into spacious new accommodation and we left the Isotope Unit in the Barnato Joel Building, with its close access to wards and other clinical facilities. This provided a most valuable stop-gap until purpose-built permanent accommodation became available.

The rapid growth of the hospital service side of the Isotope Unit in the 1950s raised another major problem. A significant proportion of the clinical tests which we were able to offer were in the *in vivo* category, requiring the presence of the patient in the testing laboratory. As head of the Unit, my great concern was that no member of the Department staff, including myself, was medically qualified to be immediately responsible, under the consultant, for the welfare of the patients when in the laboratories. Maybe my fears were exaggerated but they were real. Our long-standing close association with the Radiotherapy Department, which was also our closest neighbour, solved the problem. Quite early in the life of the Unit, Brian Windeyer allocated, on temporary loan, one of his registrars and a nursing sister, part time but later full time. At first it seemed rather a clumsy plan but in practice it worked extremely well and did much to establish the isotope work with the clinical departments of the hospital.

7.3 A New Department: The Institute of Nuclear Medicine

Rereading the foregoing paragraphs, one gains the impression of a massive exercise in improvisation and so it was. The Department of Physics had grown out of all recognition since the war and showed signs of further growth. It had become well established on two sites, with the Isotope Unit in the old building adjoining the Hospital and the main Department in the New Medical School. Towards the end of the 1950s we were beginning to think the unthinkable. Should we hive off the Isotope Unit to form a separate and independent department of the Medical School, serving the Hospital as did other departments? I agonised over this problem for many months; the idea of losing a substantial limb of the Department which we had built up since my appointment was naturally not very attractive. At the same time the idea of a new department to develop this new and exciting branch of medical science was very attractive. I soon found that this dilemma was troubling many of the major hospital physics departments in the country. It seemed at the time that most heads of physics departments favoured the idea that isotope services to clinical departments should remain under the care of physics and be provided, on request, to the clinical departments, on the lines of existing services to radiotherapy departments. On the other hand, I had come right round to the idea that nuclear medicine should be another independent branch of clinical medicine providing *in vivo* as well as *in vitro* services to other clinical departments, more on the lines of biochemistry, histology, haematology, etc, even, perhaps, physiotherapy. This would almost certainly require a medical consultant in charge. This was, I believe, distinctly a minority view at the time and many and vigorous were the arguments around the country. In my files I have a record of quite a long, hard-hitting but friendly correspondence with Harold Miller of Sheffield on this question: is nuclear medicine a branch of applied physics or a new medical speciality to be consultant supervised? We agreed to differ and acted accordingly, though I suspect echoes of this argument continue today in some places.

The proposal that the Physics Department should be subdivided and a new department of the Medical School be formed and called Nuclear Medicine was put to the Dean of the Medical School, Brian Windeyer. He received the idea enthusiastically and the plan was soon approved by the Academic Board and the School Council. During this time the wildest of ideas that the new Department should be designated The Institute of Nuclear Medicine gradually developed and, to my surprise, was eventually accepted. The new Institute would be set up as a full department of the Medical School, on exactly the same basis as the other institutes and departments. It would have a teaching programme and also a research programme orientated towards clinical problems. In addition, following long-established Middlesex practice, it would provide any clinical services in the field of nuclear medicine which may be required by the hospital staff. This transformation, wildly beyond my own original dreams, was encouraged by the fact that substantial financial support was forthcoming from the Nuffield Foundation, the Medical Research Council and the British Empire Cancer Campaign.

The new Institute was formally set up as an independent department of the School in June 1961 and I was appointed its first Director, alongside my responsibilities as Head of the Department of Physics Applied to Medicine. I accepted this with the very clear understanding that the appointment should be temporary. As I have made it clear earlier, I had no doubt that such a department should be headed by someone medically qualified. At that time, Edward Williams, qualified in physics and in medicine, was the Registrar allocated to us from Radiotherapy and he was appointed Deputy Director. When, two years later in 1963, I took a sabbatical year in the Middle East, Dr Williams took over as Acting Director. On my return in 1964, I formally resigned the post to make way for him to be appointed Medical Director. My long-term dream had come to fruition.

It would be inappropriate for me to try and continue my own meanderings into the later development of the Institute to its present prestigious level. This is another story and the first 25 years of it has been well told (1987) by Professor E S Williams and Professor P J Ell, the successive medical directors in the Institute of Nuclear Medicine. It is, however, difficult to resist the temptation to record in this context two items which started as very much the work of the Isotope Unit of the Physics Department. Both have made an impressive impact on clinical medicine, particularly in endocrinology, but have never been recorded as a product of medical physics.

7.4 Thyroid Function Tests: Immunoassay or Saturation Analysis?

In the late 1950s a large proportion of the work of the Isotope Unit was involved with the *in vitro* assay of body constituents and there was particular emphasis on thyroid and steroid hormones. The thyroid work arose naturally from the worldwide interest in thyroid function studied by using iodine-131, whereas the steroid work resulted from a more local Middlesex interest in the adrenals. As part of the clinical service of the Unit, Roger Ekins produced a series of isotope tests of thyroid function. Actual measurements of thyroid hormone levels in body fluids proved very difficult. Eventually J F Tait and J K Whitehead developed the technique of 'double isotope derivative analysis', as a very sensitive system of analysis for thyroid and steroid hormones. Although technically difficult, this proved invaluable in the hormone studies, particularly the steroids. As early as 1954, Ekins suggested a new approach to the microanalysis of these and similar substances, a technique which he designated as 'saturation analysis'. This, of course, has now been, in one form or another, a widely recommended and powerful method of microanalysis. I have neither the intention nor the ability to describe it here. In the early days, work was concentrated on the assay of serum thyroxin but the general method was eventually widely applied in other fields. Almost simultaneously with the publication of Ekins' definitive article, Berson and Yallow in the United States published a description of the technique which they called 'immunoassay', with its application to the assay of serum insulin. The two approaches were so closely related that a great controversy arose concerning details of technique and, sadly, priorities of publication. At this long distance in time, it would be most improper for me to revive such controversies, even though Nobel Prize questions were involved. I fear I might be prejudiced. Whether we think in terms of immunoassay or saturation analysis, it represents a remarkable combination of physics, biochemistry and endocrinology.

7.5 The Discovery of Aldosterone

Also in the general field of endocrinology is the second offshoot from medical physics which I must record. This was the discovery and characterisation of the major adrenal steroid, aldosterone. In the early 1950s substantial amounts of radioactive sodium and potassium became readily available. With the cooperation of Sylvia Simpson of the Courtauld Institute of Biochemistry, James Tait set up a microbioassay system for the assessment of the mineral activity of the known adrenal steroids, using the sodium-24/potassium-42 ratio in adrenalectomised rats. Having dealt with the known adrenal steroids and put them into order of mineral activity, they turned to bovine adrenal extract and found mineral activity far beyond that of the known steroids present. The search was on for a new, very highly active compound and they turned to the then relatively new techniques of chromatography. It was tempting to compare this attempted separation of the highly active component in the bovine adrenal mixture with that bit of medical physics ancient history, the isolation of radium from the uranium ores, though I fear Professor Tait would scorn such frivolous comparisons. Following a tremendous effort, the active compound was isolated and its general chemical nature defined. With its electrolytic activity in mind, we duly named it electrocortin. Some time later, the detailed structure was determined by Reichstein and others in Switzerland and, in view of that structure, the chemists overruled us and gave it the more appropriate chemical name, aldosterone. Tait and Simpson followed this with a wide range of researches in adrenal endocrinology but might be said to drift outside the more strict definition of medical physics. As a footnote to this story, James Tait and Sylvia Simpson married in 1956 and received a whole range of distinctions and awards, including Fellowships of The Royal Society.

7.6 Conclusions

In the planning of a closing paragraph for this chapter, my thoughts turned to the arguments surrounding the almost revolutionary idea of splitting off nuclear medicine as a separate academic subject. The thought was stimulated by more recent developments. Within the Institute of Nuclear Medicine, there has grown up what seemed to become a subsection of molecular endocrinology, involved with Dr Ekins' microbioassay techniques. In 1982, this unit became a full Department of Molecular Endocrinology, with Professor Ekins, an original physicist, as Head. I quote from the official history of the Institute of Nuclear Medicine:

> The recognition, almost 25 years before, that the medical use of radioactive materials merited the separation of the work from the Physics Department and the formation of a new Department, had now, in effect, been repeated by a similar recognition that radioactive tracer based microassay work was now so far reaching in research and routine diagnosis, that it in turn should continue as a Department separate from mainstream Nuclear Medicine.

Many of the other aspects of what we understood as nuclear medicine are now closely related to conventional diagnostic radiology, forming the combined discipline of clinical radiology. We have come a long way from those heady days around, say, 1948, when we just dreamed excitedly of the wonders which would result when all those man-made radioactive isotopes would become available.

CHAPTER 8

PROFESSIONAL PEAKS

The meanderings so far in these pages have been, I would suggest, very much at the ground level of medical physics. We have wandered around laboratories, hospital wards and lecture rooms, measuring, calculating and planning treatments, researching more and more abstruse problems. Already we have seen, over a period of 50 years or so, how the way in which medical physicists operate have changed and developed.

There is another level, however, in which professional people operate and perhaps meander, namely their professional organisations and societies. In these, fellow members at all levels meet informally, work is described, problems discussed, results of research and other work disseminated and published. Two such organisations have influenced and even perhaps dominated my professional life and I could not finish my domestic meanderings without paying tribute to that domination. The two are, of course, the British Institute of Radiology (BIR) and the Hospital Physicists' Association (HPA), from now on to be known for ever (I hope) as the Institute of Physics and Engineering in Medicine.

8.1 The British Institute of Radiology (BIR)

I joined the British Institute of Radiology in 1932, almost before I had found my hospital physics feet. I started work at The Cancer Hospital (Free) on 1 September 1932 and I seem to remember being hauled off to the September monthly meeting of the BIR as a first duty to my career. There I was introduced to distinguished radiologists, whose names at that time meant nothing to me, and to a few physicists whose names rather scared me. That monthly Thursday meeting of the BIR became a must for many years and I came to value it. In that little lecture theatre at 32 Welbeck Street, once a Russian Orthodox chapel, much of early

Reid-Knox Lecture Hall, reproduced from the British Institute of Radiology Year Book 1934–1935.

British radiology and radiological physics was hatched and discussed. In the earlier days, Mayneord and I were working hard on the final establishment of the röntgen in dosimetry and BIR meetings served a valuable purpose for trying out ideas.

As the membership of the BIR was something like 85 per cent medical, this meant that non-medical members had to play a somewhat disproportionate part in the organisation. It also meant that non-medical members tended to be elected to Council rather early and to spend longer periods serving. I was caught up in this problem but I think I can honestly say that I found serving on the governing body of a national organisation was a valuable part of my social and scientific education. The crunch was inevitable. Again the BIR chose its presidents year

Professor J E Roberts, President of the British Institute of Radiology 1951–1952.

by year alternating medical and non-medical. Using a sliver of simple arithmetic, a non-medical member was four or five times as likely to become President as a medical counterpart. This I welcomed as long as we realised what was happening. For me, the bell tolled or the Council called around 1951 when I had been a member for nearly 20 years. I had been a member of Council for quite a long spell and was elected to the office of President for the year 1951–1952. Even today, almost 50 years on, I can scarcely believe it happened. I was to sit in the high chair which had been graced by Sylvanus Thompson, Thurstan Holland, G W C Kaye, Alfred Barclay, Val Mayneord and many others of distinction, before and since 1952. The thought that my portrait would hang on that staircase at 32 Welbeck Street and later at Portland Place was both humbling and thrilling.

The actual year of office was far less frightening than I had anticipated. The secretarial staff of the BIR felt it their duty to shield the President from almost everything and my own secretaries at the Middlesex were most helpful. The Council treated its President with the utmost care and respect and presented no great problems. I can think of only one incident concerned with the monthly scientific meetings. Each month it was the practice to send around a voting box for the election of a group of people seeking membership of the BIR. This was very much a routine business as the candidates had already been carefully vetted. On one occasion, as the box was brought to me on the platform, I caught out of the corner of my eye a black ball in the voting drawer. This meant that the whole batch should be sent back for further investigation. This was most embarrassing and I could not believe it was genuine. On the spur of the moment, I decided to ignore the black ball and risk the consequences; the membership applications were approved and there was no protest. I have the strongest suspicion that this was a joke, not very funny, intended to embarrass me and I have an even stronger idea of the name of the distinguished radiologist who perpetrated it—just a little cruel.

Many years later, I again unexpectedly rose in the BIR hierarchy. Towards the end of my sabbatical year in the Middle East, I received an urgent message from the BIR saying that it was having some difficulty in filling the office of Non-medical Editor of the *British Journal of Radiology* (BJR) and, as I had some editorial experience, would I take the post for a time? By that time I was becoming a little long in the tooth for such an appointment but after much thought, I decided I owed it to the BIR to help. The job was never quite so frightening as I had imagined. The BJR was a very well-established international journal with very good editorial and printing services. Furthermore, the burden was reduced by the fact that non-medical papers comprised less than one-quarter of the total and were in steady supply. Organising refereeing was no problem and I rather like proof reading. I would not pretend that I contributed anything very significant to the story of the BJR but it is very satisfying in retirement occasionally to remark that I was once Editor of the *British Journal of Radiology*.

8.2 The Hospital Physicists' Association (HPA)

My experiences with the HPA and its successors were, in many ways, quite different. I have already given some account of the formation of the HPA and I was, of course, one of the 53 founder members. The organisation was governed by an Executive Committee, chaired originally, I believe, by Professor Russ. The Chairmanship was passed on annually through the senior members and my own turn came fairly quickly in 1951, just before my Presidency of the BIR. This was clearly a much more down-to-earth business than presiding over the BIR. The infant HPA, on its formation, had given itself a mammoth dual task. Its primary purpose was clearly to raise the scientific standards of physics in medical practice and the professional standards of those who practised physics in medicine. To this was added the difficult task of ensuring that the working conditions, remuneration and professional acceptance were worthy of the service they were offering to medicine. On the face of it, this sounds impossible; to put it crudely, a learned society with trade union accessories. Somehow, it worked successfully for many years. My recollection of that year as Chairman is of long weary committee meetings with an enormous agenda. At the time, London was the only practicable venue and hard-working Committee members regularly travelled long distances to keep the business moving. It was from this hum-drum, hard-working HPA that there arose perhaps the most exciting and frustrating period of my professional experience.

8.3 J E Roberts: Founding Editor of *Physics in Medicine and Biology*

Around 1955, when the HPA was still quite a young organisation, someone had the bright idea that medical physics needed its own journal. Even to today, I have never clearly understood whether the

PHYSICS in MEDICINE and BIOLOGY

A Taylor & Francis International Journal published in association with the Hospital Physicists' Association

Publishers
TAYLOR & FRANCIS LTD.
Red Lion Court, Fleet Street, London, E.C.4

The Journal will welcome contributions from all countries and will print papers in English, French or German as required, with summaries in all three languages.

Papers by authors residing in a country or region in which there is a member of the Editorial Board should send their contributions to him. Other papers should be sent direct to the Editor at the address given above. A leaflet giving detailed instructions to authors on preparation of papers for the Journal is available on request from the Editor or the Publishers.

Editorial Notice

Physics in Medicine and Biology aims to fill some of the gaps in the publication facilities available to physicists and others working on the borders of physics, biology and medicine. Emphasis is on original papers reporting results of researches in two general fields: (*a*) studies of the constitution and physical properties of living matter and its components at all levels of organization, and (*b*) applications of physics and the methods of physics to the elucidation of problems in biology, physiology and clinical medicine. In addition, occasional critical review articles on subjects of topical interest are published. Space is also made available for short notes on work in progress and on instruments of interest to workers in the fields covered by the Journal.

First announcement for the journal Physics in Medicine and Biology.

original suggestion came from the HPA or from Taylor & Francis, well-known publishers of scientific journals, including the historic *Philosophical Magazine*. The idea of its own journal was, of course, very tempting for the growing HPA, especially as the publishers were willing to cover the inevitable financial losses of the earlier stages. Again, I will never know how the appointment of Editor was decided upon. Incredibly, I was asked to take the appointment, even though I had no experience whatever of scientific publishing. I had many hesitations, but the challenge was irresistible and I accepted it. Taylor & Francis promised limited technical assistance, but editorially, I was on my own. I even had to buy myself a copy of the British Standards Institution booklet 'Printer's Proof Corrections' so that I could learn the art of manuscript and galley proof reading.

There was quite a problem over the title of the new journal. The most obvious name was 'Medical Physics', but as the prospective editor, I raised strong personal objections. I have never been happy with the idea of 'qualifying' physics. Physical chemistry I could take but not chemical physics, for example. I freely admit the lack of logic. In the end, the HPA Committee seemed to accept my point of view and took up the suggested *Physics in Medicine and Biology* (PMB), which fortunately has well stood the test of time. It would be difficult to imagine medical physics now without the monthly arrival of those prosperous-looking red-covered volumes.

The prospective birth of the journal was announced and contributions to the text invited; these came in agonisingly slowly at first and it was indeed a wretched time. The first few deadline dates were truly traumatic but, somehow, each time there was enough good material for the particular issue. At that time, I suppose at least 90 per cent of all hospital physicists were involved in radiotherapy and this was reflected in the papers offered in the early years. After four years or so, the time had come for change. Commercially, PMB was finding its feet, but editorially, even I realised that it needed an imaginative boost. We needed a new Editor and, to my surprise, found that Joe Rotblat was willing to have a go. He was duly appointed to succeed me. His energetic imagination was just what PMB needed and it has never looked back. The rest is history. When, each month, PMB arrives on my desk, I look with amazement at the range of subject matter and even more so at the enormous international range of authorship. It must now be regarded as one of the great international journals in the medical field and I am inordinately proud of my own small contribution to its origins.

8.4 Institute of Physical Sciences in Medicine (IPSM)

Much later, around the time of my retirement, the HPA, under its new learned society guise as the Institute of Physical Sciences in Medicine (IPSM, later IPEM), did me the supreme honour of making me an Honorary Fellow. This may not seem much from a new and rather small society but then I notice that the list of Honorary Fellows is remarkably short. As I went into retirement, I could not have wished for a more fitting climax to my medical physics meanderings.

8.5 The Society of Radiographers

There is, in fact, just one more organisation which deserves at least a brief mention in these meanderings. At the headquarters of the BIR at 32 Welbeck Street there was also housed the then Society of Radiographers. Even in those earlier days, the Society had the profession of radiography well organised and under control. The examinations and certificates of the Society were already recognised as a requirement for employment as a radiographer in medical radiology or radiotherapy. The courses and examinations involved included a significant physics component, and the few hospital physicists around were called upon to help with teaching and examining. The Middlesex, like many other general hospitals, had its own School of Radiography and for this our Physics Department was called upon to provide lecture courses. This raised a serious problem. Physiotherapists also required courses of physics lectures but for these we were paid, I think, £2.00 per lecture, whereas radiographers had to be taught for nothing. The problems of allocating physics lectures to staff members (fortunately this was mainly when Russ was in charge) can well be imagined.

8.6 The Stanley Melville Memorial Medal

One consequence of all this was that I got a medal. The Society of Radiographers had in its gift the Stanley Melville Memorial Medal, a

The Stanley Melville Medal.

tribute to one of the great radiologists of the earliest days and whom I remember as still active. To my great surprise, in 1954 the Society named me for this medal and I have held it since with great respect and pride.

The receipt of the Stanley Melville Medal inevitably involved the delivery of an eponymous memorial lecture. This might have seemed a bit of a burden but it offered the great compensation of a rare opportunity (for me the only one) to perform in that greatest of all scientific lecture theatres, The Royal Institution, scientific home of Michael Faraday. This I did and chose what I thought was a timely subject. The nuclear reactors were still producing new radioactive isotopes and I raised the possibility that a radioactive source might one day replace the X-ray tube target in medical radiography. This was, in fact, happening in industry but, of course, its 'patients' were liable to be inanimate steel castings and required high-energy gamma

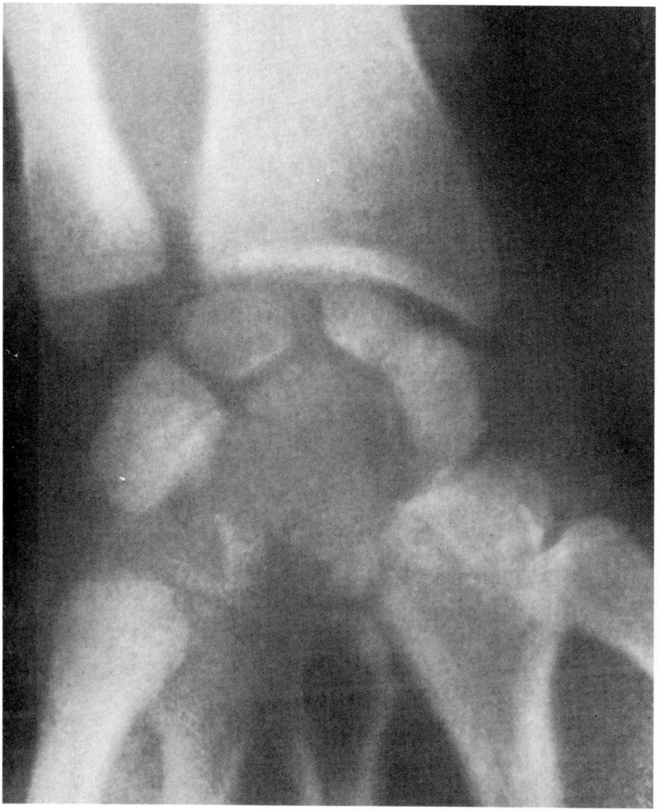

Gamma radiograph of fractured scaphoid, reproduced from J E Roberts 1956 'A gamma-radiograph' British Journal of Radiology, Volume XXIX, No 347.

rays to penetrate them. I found a nuclide which emitted radiation of about the right energy; with an intolerable exposure, I produced a quite presentable radiograph of a fractured human scaphoid. This was, in fact, the radiograph mentioned in Chapter 10 as of the quality of a 1930s' X-radiograph. As one might expect, the exposure was very long but I had made my point. Unfortunately, since that time no nuclide has appeared with the right range of radiation energies coupled with adequate specific activity and we must regard medical isotope radiography as a permanently lost cause.

8.7 A Lecture at the Royal Institution

Lecturing at the Royal Institution has one most peculiar consequence. There is a very long tradition that the lecturer, however distinguished or otherwise, is locked alone in the anteroom for 10 or 15 minutes before the steward opens the connecting door to the lecture theatre. Apparently legend has it that in Faraday's day one potential lecturer chickened out and ran away whilst waiting for his performance. The locking up order was given and has never been revoked. I count it a privilege to be included in the quite historical list of scientists who have followed this strange procedure. I count it as one of the very unusual meanderings in medical physics.

CHAPTER 9

PHYSICIST—THE ODD-JOB MAN

I suggest that this chapter should be regarded as a comfort break between my meandering through the growth of medical physics in Britain and wider meandering observing and encouraging the different ways in which medical physics has taken root in other countries round the world.

In the early days of the profession, the hospital physicist worked in a very narrow field compared with present-day counterparts. The few who worked in teaching hospitals taught some very basic physics to medical students and perhaps basic radiation physics to potential radiologists and radiographers. In the hospital itself, work was usually confined quite rigidly to the clinical applications of X-rays and radium. It was all so different from the present time when physics and engineering have infiltrated and in some cases become significant partners in a variety of clinical fields.

However, occasionally technical problems arose which puzzled or defeated the hospital administrators or clinicians.

In those halcyon days, hospitals were administered by the Medical Committee and maybe the Matron. Most of these problems were trivial but now and then I can hear some consultant or secretary saying 'we have a Physics Department, why not use it?' Again most of the problems are forgotten but I find that I can recall a small number. Recall is the operative word as records are generally non-existent. Perhaps reviving a few of these may stimulate some hospital physicist to make a study of 'odd jobs for physicists'.

9.1 Electric Shock Therapy

The first odd job which comes to mind arose from a lunchtime conversation with the Consultant in Psychological Medicine. I clearly recollect that his approach was something to this effect: 'So far I have only asked you questions about gardening but now I have a real problem.' I must not encroach on the clinical problems, but this was at a time when electric shock therapy for some psychiatric conditions was booming, though even then it had its critics. The problem posed to me was that of timing the shocks to be administered to the patients. These were normally prescribed between 0.1 and 1.0 seconds. Any system proposed had to be very simple, cheap and effectively foolproof. The very surprising answer we suggested proved remarkably simple, accurate and unexpected. I am fairly certain it came originally from the fertile imagination of Frank Farmer. The answer we suggested was in the form of a common or garden dial telephone. We found that if one dialled zero, the instrument took just 1 second to return to base. Moreover, dialling 9, 8, . . . , 1 gave timings of 0.9, 0.8, . . . , 0.1 seconds, just what the doctor ordered. There was no difficulty in wiring the dial phone into the main shock equipment to provide the required range of shock times. I understand the instrument worked quite well in practice, but this was wartime and the idea was not carried much further. Perhaps I should draw a veil over how we came to have a spare telephone dial unconnected to the main telephone system and strictly the property of the Post Office. This was, of course, before the privatisation of telephones.

9.2 Anaesthesia

My second case of unusual assignments (odd jobs) could not, I fear, be treated so light-heartedly. In April 1952, there was a serious accident in one of the main operating theatres at the Middlesex, as a consequence of which a patient died. Apparently, just as a major operation was about to commence, there was a large internal explosion in the anaesthetic apparatus, for which there was no immediately obvious cause. The Medical Committee asked me, with the help of one of the consultant anaesthetists, to investigate the matter and report back. From the physics point of view, this was a complete shot in the dark as I had little or no knowledge of anaesthetic techniques or equipment. At the time of the explosion the patient was fully anaesthetised, the anaesthetic being a standard cyclopropane–oxygen mixture in a closed circuit which, of course, included the patient's lungs. Damage to the lungs was very severe and the patient died a few days later. As these notes are intended to have a historical flavour, it is interesting to note that in our eventual report on this accident, there is a reference to the 'increasing use of ignitable anaesthetic gases', whereas to show the report to a present-day anaesthetist, as I have done, provokes just a wry smile of disbelief. There was no naked light and no live electrical equipment near the operating table. We eventually came to the conclusion that the most probable cause of the explosion was an electrostatic spark near a leak-off valve which could have been momentarily and accidentally opened by movement of the respiration bag. The obvious recommendation was that every effort should be made to eliminate all possible sources of electrostatic charge anywhere in the theatre. We suggested the replacement of all wool and plastic fabrics, such as bed covers, by cotton, a poor producer of static electricity. All rubber attachments in equipment, trolleys, footwear and so on should be of the carbon-loaded conducting type. Somewhat later we were able to introduce conducting floors to the theatres. Most of our suggestions were readily accepted by the Medical Committee and I believe many of the ideas came into widespread use elsewhere. This event even lead to a paper in the specialist journal *Anaesthesia*, a very rare experience for a working hospital physicist. Yet it was surely physics at its most basic. Such is the progress of our subject that it is not unknown today for physicists to be employed full time in departments of anaesthesia.

9.3 Radiography of Postage Stamps

Another odd job which I recollect had little or nothing to do with medicine, except in that it involved the use of equipment which was clearly medical. One of my Medical School colleagues, Mr W H Cheavin, was also a philatelist of some distinction. Although he tried hard, he never succeeded in recruiting me into that fraternity. However, from time to time he posed technical questions which had to have answers in physics. As in any other branch of society, there are clearly some villains in philately, people, for example, who forge rare stamps or those who tamper with and repair valuable damaged stamps. I was asked, for instance, as to why we could not 'X-ray' stamps, to detect forgeries and repairs. The very idea of radiographing paper seemed like nonsense until a very remote possibility occurred to us. The answer seemed to lie in the Radiotherapy Department rather than in radiodiagnosis. At that time we were using an X-ray tube, working at 10 kV for the treatment of very superficial skin lesions and occasionally for clinical depilations. This looked a real possibility for the radiography of postage stamps. With appropriate permissions, I tried it and it worked beautifully; it was particularly good in sorting out inks, especially the older metallic varieties from the modern organics which could be used in forgeries and repairs. It also demonstrated differences in paper constitution so that many other forms of forgery and repair were covered. We even radiographed a few stamps from the royal collection. An interesting scientific alternative arose, namely the possibility of using 200 kV X-rays for some stamp radiography. A method of radiographing fabrics and paper, using secondary electrons, had been reported. We went a step further. Using older stamps with metal-based inks, we irradiated them with 200 kV X-rays and the secondary electrons emitted from the ink metals produced excellent positive images of the stamp pictures. This was all a fascinating exercise which I fear was not seriously followed up. Philately may have missed a great opportunity; for me, it produced a letter in *Nature*. This must be regarded as a one-off oddball reminiscence. I could not pretend it is really meandering in medical physics but perhaps hospital physics, as the equipment is only likely to be found in a hospital

9.4 Lighting Engineer

Soon after the end of the Second World War there was a move to rebuild much of the Middlesex Hospital Medical School. One of the first projects to be completed was the very fine Boldero Library. The main reading room followed the well-established pattern of bays with bookracks and reading tables. After some time in use, a major defect was suspected: the ambient lighting at the reading tables was inadequate and badly distributed. The Medical School Council discussed the matter and came to the conclusion that, being concerned with light, this must be a physics problem and that the School had a Department of Physics. My first reaction was fiercely negative but I found it very difficult to maintain this position. I eventually accepted the challenge and promised a technical report. I realised that there was a profession of ‘lighting engineering' but was vague on its principles and methods and decided to soldier on. Scientists working in central London have one great advantage, namely the Patents Office Library. Here, with patience, it was possible to 'mug up' on almost any conceivable branch of applied science. This is exactly what I did; I became for a short season a self-taught lighting engineer. In the laboratory there was no difficulty in finding instruments for measuring light intensity. More dramatic, however, was the fact that I was studying the distribution of radiation on a plane surface from a distribution of linear sources. I was almost inclined to look around for my Paterson–Parker tables; the similarity of the problems to those of radium applications was dramatic. The end result had much in common with a radium surface applicator plan of the period. We had a distribution of strip lamps which should produce an adequately intense and uniform light on the library tables. The report must have been impressive as the Council readily accepted it and put it into production. In all modesty, I believe it proved a complete success. In the light of the breathtaking modern developments in our professional work, this may not be technically outstanding, but it is still probably a unique bit of meandering in medical physics.

CHAPTER 10

ICR, ZURICH, 1934 AND ICR, CHICAGO, 1937

10.1 The 4th International Congress of Radiology

In the 1930s, overseas travel was somewhat more difficult and time consuming and also much less freely available than today. As a consequence, there were far fewer international scientific and medical conferences, and attendance tended to be limited to the more senior members of the professions. Thus it was that I had to consider myself extremely fortunate to be allowed and encouraged to attend the 4th International Congress of Radiology in Zurich, in 1934, being at the end of only my second year as a hospital physicist. As we have seen, radiology was the only medical field in which physicists were directly engaged, making the Congress an important point of contact for the small number of us involved. It seemed that a few comments and anecdotes from this almost forgotten occasion might add interest to the meanderings of myself and others in this field. I fear it may also be the occasion for some name dropping and I offer appropriate apologies.

The figure 4 in the title of the Congress is, in itself, quite remarkable. It all happened less than 40 years after the discovery of X-rays. Communications were orders of magnitude slower than today but radiology had become a wide-ranging world speciality, with professional organisations in many countries. In that year, 1934, 33 such organisations were represented at Zurich, led by the oldest of them all, the BIR, which had taken over from the original Rontgen Society. Yet this was the fourth international congress, the first being in London in 1925. The examples of radiography and radiotherapy which were demonstrated might well have been very primitive by today's standards but the devotion and effort which went into their production were most impressive and sometimes heroic. Many years later when a variety of radioactive isotopes had become available, I remember showing with some pride, to a radiologist colleague, a simple radiograph of a bone fracture which I had produced with a radioactive source instead of an X-ray tube. His comment was instant and interesting: 'That looks very like a radiograph from the early 1930s'. I was rather proud of that, though the idea of radionuclide medical radiography never took off; X-rays, followed by ultrasound and MRI, hung on to the market. It would be of little use in this context to attempt an analysis of the technical, scientific and clinical content of that far-away Congress. All I would offer are personal reminiscences and comments on the wider aspects of the Congress and the people who met there.

The records show that the total registered membership of the Congress was just under 1000, a fraction of the size of present-day meetings. The nationality make-up of the meeting is most interesting. Of the 1000 members, only 52 were from the United States, an almost incredibly small figure, with only some 63 from Great Britain. Continental Europe, where, after all, radiology was invented, dominated the membership, with 186 from Germany, 150 from France, 120 from Switzerland and about 40 from elsewhere. One curiosity is quickly noted. China claims one representative but he turns out to be the obviously very English Superintendent of the County Hospital in Shanghai. Just looking through the portrait catalogue of members of the Congress is, for me, a fascinating experience. One has to realise that many of the people appearing there were adult and some practising medicine when X-rays were discovered in 1895. On almost every page famous names and faces stand out. And I was there, very much the new boy, mixing with them, talking with them and learning from them, the legendary figures in the history of British and world radiology.

10.2 The Journey to Zurich

The process of getting to Zurich presented problems and, in fact, for some of us provided most interesting experiences. I suppose, as a very junior member of the establishment, I should not have been there at all; vanity suggests it was a fair reward for much hard work. To make it more exciting and just a little frightening, as Mayneord and I had made some progress with the intractable problem of measuring gamma-ray

Dr Tchaperov, J E Roberts and W V Mayneord at the 4th International Congress of Radiology, Zurich, 1934.

dose, I was to be allowed a direct share in presenting this work to the Congress.

10.3 Courtesy of Philips

Before we come to all that, our transportation to the Congress would almost provide a story in itself. There were, of course, no such things as comfortable flights by British Airways or Swissair from Heathrow to Zurich. Heathrow was still covered by market gardens—I lived nearby. The best available route involved a Channel crossing followed by a long train journey across Europe. For some people this prospect was changed when the BIR received a surprise offer from Philips of Eindhoven, then, as now, a major producer of radiological equipment. The company offered to provide a coach at the Hook of Holland which, over a period of three days, would transport a party of British radiologists and others to Zurich. The only condition appeared to be that the party would spend half a day visiting the Philips factories in Eindhoven. Other professional and sightseeing visits were also promised. On the evening of 20 July the BIR party of some 30 radiologists and two or three physicists sailed out of Harwich to be met by the Philips coach at the Hook of Holland next morning. I was a very wide-eyed member of that party and it is a sobering thought to realise that, although I was a graduate of some six years standing and a PhD for four years, this was the first time I had set foot outside Britain. Long-distance coach travel was then quite uncommon and the long haul along the Rhine and through the Black Forest was an almost unique experience for most of us.

On the first morning the party was entertained with full Dutch hospitality at Eindhoven. We were instructed too. Philips was deeply involved in the 1930s' transformation of radiology from its DIY state to a more recognisably modern form. Philips was introducing the Metallix tube with its earthed centre, the forerunner of most modern tubes and, together with Siemens and a number of smaller firms, it was changing the face of departments of radiology. The visit must have raised a number of radiological eyebrows and did no harm to Philips' business prospects.

10.4 Via Cologne (Köln)

Our first overnight stop was Cologne. Memories of that youthful trek have returned to me on a number of occasions in the past 60-odd years when I have returned to that great city and looked up at those cathedral spires. Visits included Dr Rudolf Grashy's X-ray Department at the Burger Hospital in Cologne and Dr Robert Janker's department in Bonn. Among the group that visited the Burger Hospital were quite a number of characters who have played an important part in the history of British and international radiology: Ralston Paterson, who is almost synonymous with the Christie Hospital, and the diagnosticians Eric Duff, also from Manchester, and Richard Connell of Carlisle; Duncan White, President in that year of the BIR, probably the last of the pioneer X-ray martyrs, a brilliant radiologist and a very brave man; and Alfred Barclay, one of the greats of British and world radiology.

To digress for a moment, whenever Barclay's name is mentioned, my mind springs into vigorous action. Many years later, during my own turn as President of the BIR, the Barclay Medal was founded in his memory. During that year the Council, over which I was supposed to preside, spent innumerable hours discussing the design of the Barclay Medal. The most intractable point at issue was whether the figure of Barclay on the medal should be shown with or without his pipe. The Council eventually decided that it was not really competent to pronounce on matters artistic and left the final design to an individual, not the President.

Returning to our Cologne group, among the English party I recognise one Dutchman, Albert Bouwers, presumably representing Philips. A brilliant X-ray engineer, he played a most important part in the technical transformation of radiology in the 1930s. In such surroundings, I was flattered to find myself, a youngster of only two years' professional standing. Two other young physicists were also present, Tom Banks of Bart's, and Tom Chalmers, then also of Bart's but later to find fame in Liverpool.

There were other professional visits on this journey to Zurich, for example at Frankfurt (for me this meant the Holfelder approach to planned radiotherapy) and Freiburg. I remember one interruption of the journey which has nothing to do with radiology nor medical physics. As a result of hospital visits, we were running very late one evening and a drink and 'comfort' stop in a Bavarian village became imperative. In the village hall a dance was in progress, to which we were immediately invited. Such is the selectivity of memory that, to this day, I have a clear picture of several of the pillars of British radiology, such as Alfred Barclay and G W C Kaye of the National Physical Laboratory, tripping the light fantastic with the village maidens. Sadly, this memory contrasts with other rather frightening recollections of marching, shouting Nazi Hitler Youth members in some of the places visited.

The Barclay Medal.

10.5 An Expensive Bath in Zurich

After three days of hard travel (three hours is more usual today), our party arrived in the Congress city, Zurich. That arrival again evokes memories, and an irrelevant and unflattering incident again breaks the seriousness of medical physics. In those days, 'en suite facilities', even in superior hotels, were rare and, on arrival, I needed a bath. Indeed I got my bath and very good it was, too, but I still remember my horror when I found that I was to be charged 10 shillings (50p), maybe the equivalent of £10 or £15 today for just half-an-hour's immersion. That 50p changed my ablutionary practices for the next fortnight.

10.6 The Congress

This, my first trip abroad and my first taste of international congresses, was an experience never to be forgotten. Of course, in the Congress sessions I was extending my knowledge and understanding of the expanding field of radiology and its physical ramifications. Much more important was meeting the people concerned, meeting them over breakfast tables or in hotel lounges and lecture halls. I was right alongside the British contingent, normally seen only occasionally at the BIR. There were characters in the process of becoming part of the history of British radiology: there was Cuthbert Andrews, doyen of the commercial world of X-rays, Russel Reynolds, Charles Thurstan Holland, Richard Fawcett, E W Twining, or in radiotherapy, Stanford Cade, N S Finzi, Walter Levitt, lawyer and radiotherapist, Constance Wood and many others. Such a list is woefully discriminatory. On the wider canvas of world radiology, the picture was just as impressive. I was meeting informally people with names I had heard mentioned with awe in wards and treatment rooms at home, names which were essential reading in *Acta Radiologica*, *Strahlentherapie*, the *American Journal of Roentgenology* and so on. There they were in the flesh: Gosta Forrsell, Antoine Beclere, Benjamin Orndorf, Douglas Quick, Henri Coutard, Claude Regaud, Maurice Lenz and many others. Again it was difficult to realise that some of them were practising medicine when X-rays were discovered. Our new profession of medical physics, limited at that time to radiological physics, was surprisingly well represented. From Britain, in addition to Tom Banks and Tom Chalmers, I remember Hal Gray, Val Mayneord and Charles Wilson; from other parts of the world there were

Gioacchino Failla and Edith Quimby from New York, Lauriston Taylor from Washington, DC, Otto Glasser from Cleveland, Walter Friedrich from Berlin and, of course, Rolf Sievert from Stockholm. That really must be the end of name dropping.

10.7 The Scientific Programme

I have neither the knowledge nor the expertise to attempt any general account of the clinical or technical content of the Congress programme. I must limit myself to non-expert generalisation and a few items which we may now regard as curiosities. The programme of the Congress was planned around a series of some 10 keynote papers by recognised specialists in the fields involved. Although techniques have changed almost beyond recognition, most of these would be clearly recognised today. I picked out two themes which would be most unlikely to appear in a modern radiological meeting.

We were reminded that X-rays did not start life as a medical but as a physical phenomenon and that even in 1934 they had many non-medical applications. A distinguished non-medical member of the Congress, Dr W T Astbury, from my own University of Leeds, and one of Britain's greatest X-ray crystallographers, gave one of the review lectures on the structural analysis of biological materials, a subject which he pioneered. From this type of work, the way was opened years later for the revelation of the DNA double helix, a subject not normally associated with radiology nor, perhaps, with medical physics.

The second curiosity must seem almost out of this world to a radiologist or physicist in the 1990s. It was a review paper by S Salkind of Leningrad (I should now write St Petersburg) on mitogenetic radiations, those strange emissions which it was claimed arose directly from biological reactions at cellular or even molecular level. It was easy to raise a superior smile at the mention of these phenomena but they were taken very seriously by some reputable scientists, including one or two medical physicists. Not surprisingly, many people did question the existence of these radiations, but they persisted in the literature for some years before quietly disappearing from spectrum and journals.

In the main body of presentations to the Congress there were some 435 papers, which by present-day standards would be considered very thin on the ground. There were 156 in radiodiagnosis, 115 in radiotherapy, 39 in radiobiology and 50 in technology and physics, perhaps a not unreasonable balance. The remaining 75 would probably be scorned by a modern radiological organisation. There was quite a large separate section of 58 papers devoted to electrology, papers which would undoubtedly today be shuffled over to physiotherapy. We might remember, however, that in the 1930s, before the Royal Colleges took over the radiological diplomas, the Cambridge Diploma was quite specifically DMRE (Diploma in Medical Radiology and Electrology) involving some knowledge of electrology, which, I can assure the reader, posed some problems for the physicists. In the Congress programme there was also a small section on heliotherapy, mainly from French and Swiss authors, covering the therapeutic uses of natural sunlight, ultraviolet and infrared radiations.

10.8 The British Contribution

The British contribution to all this was quite surprising and just a little disappointing. Of the 156 papers in radiodiagnosis, only two were from Britain and in radiotherapy there were also two out of 115 and neither of these was from a clinical radiotherapist. Britain also contributed two papers in radiobiology and three in physics, two of these being from Mayneord and myself. To be quite fair, our normally prolific American friends produced only eight papers in diagnosis and nine in therapy, not quite up to their present-day standards. We might well regard the content of the programmes as somewhat primitive but it was reflecting the rapid modernisation of equipment and technique which was taking place in diagnostic and therapy departments. It is a little surprising to realise that ‘kymography’ was the ‘in’ word at the time, though I have an uncomfortable feeling that I did not realise what was going on. The therapists were very much concerned with matters of dosage, with which I could sympathise. It was a long way from tomography, computed scanning, interventional radiology, digital subtraction, ultrasound scanning, nuclear magnetic resonance, cobalt units, linear accelerators and so on, but we were on the way. There were glimpses of the future, particularly in therapy and physics. Failla enthused about an induction coil machine which he had coaxed up to 800 kV; Mayneord talked about the merits of telecurie units containing enormous active sources up to 5 curies, though the term telecurie unit had not yet been invented.

10.9 My First International Paper

It is interesting to note that more than one-third of the papers in radiotherapy were concerned with the treatment of non-malignant conditions. In the technology and physics sections there were papers concerned with the rapid contemporary changes in equipment development; there was some concentration on dosimetry, particularly of gamma rays, which was not surprising as the use of the röntgen was still in its infancy. It is at this point that I come to my most vivid personal recollection of the 1934 Congress. At The Cancer Hospital we had made some progress on the problem of gamma-ray measurement and I was being permitted the honour of presenting the definitive paper, which we thought to be important. It was a stressful and eventually petrifying experience. My first ever international paper almost ended in personal disgrace. At that Congress, there had been introduced for the first time the dreaded so-called traffic light system, by which the Chairman controlled the timing of lecturers. On that day, to my utter shame and dismay, I overran the red light and had to bring my presentation to an abrupt end, something I do not think I have ever done since in a very long career of speaking and lecturing. With hindsight I realise it was a trivial failure on my part but at the time it was most distressing. Fortunately the content of the paper was clearly better than the presentation and it became an accepted part of the international understanding of gamma-ray dosimetry. From this 1934 context I must offer just one quotation from a presentation which actually has little or nothing to do with radiology or physics: 'Et nous avons aussi commence des recherches sur l'homotransplantation du coeur a cause de l'importance de ce problem'. I hasten to add that the hearts in question were those of dogs.

10.10 St Moritz

Zurich did not, in fact, see the end of this 4th International Congress of Radiology. This was Switzerland and, apart from their many other virtues, the Swiss are master hoteliers and restaurateurs, with an eye for profitable hospitality. At the end of the week in Zurich, the whole Congress—delegates, hosts, administrators, exhibition and everything else—was uprooted and transferred to St Moritz for another week. As multiple parallel lecture sessions were then virtually unknown, there were evidently enough lectures available, though I suspect that many of these were presented somewhat reluctantly.

The social side of the Congress thrived and it was a great joy to meet more people amidst the delights of St Moritz. For me, this was illustrated in a number of snapshots which, for the sake of posterity, have been deposited with the Radiology History and Heritage Trust archive. My favourite by far is a picture of Charles Thurston Holland, in the full Edwardian regalia he favoured, outside the Congress hall. He was a guest of honour at this Congress, one of the earliest of British radiologists and President of the 1st International Congress in 1925. A great radiologist and an entertaining character, Thursty was perfectly happy to be encouraging and discussing life and radiology with a young upstart physicist who was radiologically wet behind the ears. Then there is a picture of him with another great radiologist, J M Woodburn Morrison of The Cancer Hospital, with, for no obvious reason, myself again just visible in one corner. Yet another of physicist interest shows W V Mayneord and J E Roberts, very young, very slim and trim, as presumably they were over 60 years ago.

10.11 The Return Home

The journey home represents part of the meanderings, and it had its moments, though somewhat different from the outward journey. Some four of the younger British representatives (no names) travelled across Europe by train. None of us, even myself on the munificent salary of £350 per annum, could afford anything better than the old, hard, wooden seats of continental rail travel; was it third or fourth class?

Though it has no bearing on medical physics, I clearly remember the long wait to change trains in Basle, the search for somewhere economical to eat, opening the door of a restaurant-cum-bierkeller and being driven back by the garlic. The journey was undeniably uncomfortable; two of us, one now a very aged and internationally respected radiotherapist and myself, were forcibly evicted from a neighbouring first-class compartment which we had claimed for a bedroom. I have attended many international conferences and meetings of various kinds since 1934, but memories of that Swiss adventure are very special.

10.12 The 5th International Congress of Radiology

Whilst I was writing up the Zurich Congress, I unexpectedly came across a surprising folder amongst the mass of old paperwork in my study. In fact the folder contained mainly a series of menu cards for transatlantic liner meals, but there was also a small pack of faintly pencilled notes, almost indecipherable. These turned out to refer to my very first visit to the United States in 1937 for the 5th International Congress of Radiology. This was still early history for radiology and for medical physics and so an important part of my own meanderings, deserving at least a brief mention here. Unfortunately the archival material in my find was incomplete. I know that I made some scientific notes and comments on the Congress and other visits but they have long disappeared. We are left with the barely decipherable words of my personal diary, notes and comments.

Before embarking on the personal reminiscences of the journeyings and meetings, we should perhaps put the 5th International Congress into perspective. The decision to accept the invitation to the United States was taken in Zurich with Chicago as the chosen venue and Dr Arthur Christie, a great name in American radiology, as President. The Americans had suggested a June date but the International Committee insisted on September. The idea of a Congress so soon after Labour Day must have raised some eyebrows in the United States. In an inconsequential note in the official Congress memorandum, I find that the 4th Congress (Zurich) passed on the munificent sum of $200 to help defray the preliminary expenses of the Chicago meeting. Compared with the 1990s, they must have been living in a different world.

In writing of the 4th Congress, I included a few statistics of membership and attendance and a comparison with Chicago is most interesting. The total registered membership was around 1300 as against 950, not a spectacular increase, but this could well have been due to the distances involved. Global travel was nothing like so simple as today, nor so widely available. This point is well illustrated by the distribution of nationalities. I quote just a few figures. Australia, Egypt and China had 2 each, England 33 (including 7 physicists), Scotland 3, Ireland 4, Germany 64, France 24, Italy 23 and the rest of Europe about 80. To these we add 24 from nearby Canada and 950 from the United States.

This was almost the same as the total membership in Zurich, with some 100 from New York City alone.

10.13 The Palmer House Hotel, Chicago

The venue for the Congress was Palmer House, then Chicago's leading hotel and in the 1930s almost unique. We are accustomed today to large international conference centres, with an attached hotel, but not 60 years ago. Palmer House could house the whole Congress, meetings, lectures, exhibitions and, at the same time, provide living accommodation for all the delegates. This was unheard of, but true. One Congress memorandum boasted that six weeks before the opening, over 1000 room reservations had been received. It was clearly good for business. My notes and recollections of arriving at Palmer House for the Congress seem to introduce a rather less euphoric note. In that diary my innocent visions of American efficiency disappeared amidst muddle and trouble, with horrendous queues for registration and information.

10.14 Chicago and Notes from the Congress

In spite of it all, there can be no doubt that Chicago was a most exciting place for an untravelled young provincial Englishman. Although alcohol had been back for some time, there was still a residue of the post-prohibition atmosphere and people were still ready and willing to talk about the dubious excitements of the Al Capone gangster era. It would appear that I took rather a lot of time out from the Congress to go visiting and sightseeing in Chicago. There are notes of a long visit to the famous Wheat Pits of the Chicago Board of Trade, surely of dubious significance for a medical physicist! There was a guided tour of the fascinating and revolting abbatoirs of the Chicago stockyards and an afternoon at a 'double header' baseball game (Chicago White Sox versus Detroit Tigers).

Diary notes and recollections of the Congress itself are somewhat mixed and sometimes vague. I was growing up professionally and the excitement of meeting friends and acquaintances in such an international context was tremendous. I am sure I attended a reasonable number of Congress sessions, particularly those involved with physics directly or in its medical applications. The general impression was that there had not been much in the way of dramatic development in the field since Zurich 1934. This did not diminish the value of endless conversations, discussions and arguments in the hotel lobbies, anterooms or the Coffee Shop. I noticed one peculiarity for a congress of this type. Some of the major lectures were put on as evening showpieces after dinner, with mixed success. One day my diary says, 'Lecture (1.5 hours on Ca Larynx), which I did not attend, followed by reception and dance—very dull'. This might be offset by the following evening's note: 'dinner, followed by Pfahler on Ca breast—a fine piece of work'. Unfortunately, I had to add 'lecture only two hours late'. I trust my American friends will forgive this raising of ghosts; the whole was nothing like so bad as this sounds. As has been indicated, the long-term recollection is of no great impression of important scientific advance, but rather of consolidation or perhaps I was becoming more sophisticated? Nothing, however, could take away the excitement of meeting people, sharing ideas and effectively becoming part of a new establishment.

10.15 American Medical Physics

Apart from the Congress, Chicago 1937 served quite another purpose for me; it gave me an opportunity to see, if only briefly, the rapidly growing field of medical physics in the new American context. Only a short visit was possible, no more than four weeks, and this included two Atlantic sea crossings. The fact that I was able to go at all was something of a mystery which was never solved. From indirect information, I was virtually certain that a relatively obscure member of the Medical School staff provided most of the necessary funds but demanded anonymity. The secret was so well kept that even after 60 years, I have never been able to offer thanks. The actual journey to the United States was something rather special. The *Queen Mary* was very new and it was the done thing to sail in her to New York and so I did just that. True, for me it meant travelling third class (we called it steerage in those days) but it was a tremendous adventure, though the crossing was a bit rough.

10.16 New York

In order to acclimatise, I had allowed myself three days in New York and, being my first visit, it is not surprising that my diary entries tend to the ecstatic. The generous hospitality of the American radiological organisations was very soon evident. Visitors to New York from Europe were asked to register in that city and we were royally welcomed. One day we were given the full works of entertainment at the Radio City Music Hall, then described as the largest stage and theatre in the world. On the following day there was a long conducted tour of New York City, when we were shown parts of that city which, for most of us, were just scenes from movie pictures. I was fortunate in that I had acquired several friends involved in radiotherapy and physics in New York and so some brief hospital visits were possible. One very much up and coming radiotherapist was Milton Friedman and he was a tremendous help. At the Belle View Hospital he proudly demonstrated his system of treatments carefully planned physically as individual problems, the very things we were trying to develop at home. Incidentally, it was at that time that Milton Friedman tried very hard to persuade Herbert Parker and myself to emigrate to the United States and swell the growing ranks of hospital physicists there. It was not an easy temptation to resist. I was able to visit Gioacchino Failla and Edith Quimby at the Memorial Hospital, perhaps the earliest of the well-known 'duos' of medical physics. I am quite sure Failla would have had a good laugh if he had seen the first comment on this visit in my diary: 'there is an awful lot of home-made equipment'. This may have been true but it included a valiant attempt at a home-made, high-frequency, million volt X-ray generator. At the same time, I commented on a wonderful instrument workshop.

10.17 Washington, DC

The journey from New York to Chicago was somewhat roundabout. For what today involves a flight of about two hours, I took trains, via Washington, with a couple of days' stopover in that city. This was an indulgence, but surely Washington is, or was, a city not to be missed, as my diary seems to indicate. Even today it seems incredible that I fitted in the White House, the Capitol, the Library of Congress, the Washington and Lincoln Memorials, Arlington, Annapolis and several other places within 48 hours. My clearest memory of that long ago visit to Washington is that in the garden of the city-centre hotel in which I stayed, I saw more fireflies than I have ever seen in my life before or since. The professional reason (excuse?) for travelling via

Washington was to visit the Radiation Laboratories of the National Bureau of Standards. These were under the direction of the great Lauriston Taylor, who had played a leading part in the development of precision X-ray dosimetry in the 1930s. Not surprisingly, the work on dosimetry was outstanding, particularly the high-precision dosemeters and the American National Standard. I think I must have smiled behind my hand as I wrote the following in my diary: 'they are working on continuously evacuated X-ray tubes up to 500 kV, but were having serious trouble with the vacuum'. The reason for my smile was simple: in Britain at that time, Metropolitan Vickers was installing a number of very successful, continuously evacuated X-ray therapy machines on a commercial basis. After two days, I took to the rail again on the Liberty Ltd for Chicago and the Congress, my second taste of American railways of the 1930s. My comment at the time was that American rail travel was really luxurious but only if you were fortunate enough to travel first or Pullman class.

10.18 Cleveland

At the end of the Chicago Congress, it had been my intention to make a quick dash to the West Coast, where things such as betatrons and linear electron accelerators were just beginning to appear and which might have relevance in medical physics. This was no quick Jumbo airliner hop but involved long train journeys, two days out and three back; before the Congress ended, I had chickened out. Maybe I was homesick or exhausted. I just planned a leisurely return home via Cleveland, New York, Boston and Canada. Apart from a look at Niagara, returning via Cleveland gave an opportunity to take up an invitation to visit Otto Glasser, one of the real pioneers of medical physics in the 1920s and early 1930s. Glasser was deeply involved in the setting up of the röntgen and at one stage he even had a recognised unit of his own. He made considerable contributions to the eventual agreement. At the time of my visit, Glasser was much concerned about the reality and the properties of the so-called 'mitogenetic radiations', allegedly produced by some biological and chemical reactions. I never really understood what was behind these strange phenomena, which eventually quietly disappeared from the agenda of medical physics. I would record, without further comment, one strange, brief sentence from my diary after this visit: 'reaction of ferrous sulphate and potassium

Dr Otto Glasser, pioneer of medical physics.

bichromate seems to produce some radiation which is detectable in a Geiger counter'. I would be interested in the comments of my 1990s' medical physics colleagues on this.

This variation in my return brought a non-scientific bonus: the rail trip from Cleveland to New York made possible a side trip to Niagara. Although this may nowadays seem fairly commonplace, in the 1930s it was something about which a naïve young man from Europe, like myself, could boast. It was an incredible experience; I even walked over to Canada for dinner. Although I have never been anything of a photographer, some time later I even won a prize in a local competition for my picture of Niagara by floodlight. This journey also raised another non-scientific curiosity in my archives. For a small part of the rail trip, I had to travel 'coach' class instead of Pullman, to which my ticket applied. Although I realised the difference might be trivial, the conductor was most disturbed and insisted that, on my behalf, he should apply for a refund. Some three months later, in London, I received a letter agreeing the refund and enclosing a cheque for 15 cents. Nearly 60 years later I still have that cheque, as yet uncashed, and have wondered what the accountants of the New York Central Railroad Company did with the 15 cent discrepancy in their accounts. It might be desirable to draw a veil over the extra 24 hours I had to spend in New York. The American Legion Convention was in full swing and the whole place was quite mad. I have to remind myself as I write that this was the meeting of the veterans of the First World War. Getting around was almost impossible, though I see that I managed a visit to the Presbyterian Hospital.

10.19 Boston

It was a great relief to move on to the relative peace and the New England atmosphere of Boston, where I was able to meet up with old colleagues from Cancer Hospital days, Dr and Mrs Brues. There was a medical physics bonus in this extra visit. I was able to go to the Huntingdon Hospital where there was the daddy of all van de Graaf machines actually treating patients at 1000 to 1200 kV. This was a first for me and very exciting. I was also able to look in on the wonderful facilities of the Massachusetts Institute of Technology, where all this began and where development continued on the van de Graaf machine. At that time, this looked very much like the way ahead for high-energy radiotherapy but it was, of course, overtaken by the Cockcroft and Walton machines and later by the betatron and the linear accelerator. I cannot resist copying what now looks like a rather pathetic comment from my Boston visit diary: 'must read this up and try it out if possible'. With our sophisticated multimillion volt accelerators for X-ray and electron therapy, perhaps we should do well to remember that the pioneers in this field used glorified Wimshurst machines, though I suspect that in 1937, the young men at MIT would not have been very happy with that description.

From Boston, the journey home had no scientific content and was uneventful, via Montreal, Quebec and a pleasant transatlantic voyage in the *Empress of Britain*, a ship which some two or three years later was sunk in the Battle of the Atlantic. I now quote only the last words of my diary of this tour, written one night on the *Empress*: 'Spent the whole evening teaching W V Mayneord and Mrs M how to play dominoes'. Meandering in medical physics does take some strange turns.

10.20 Conclusions

Since the war, there have been regular International Congresses, including London, Munich, Mexico, Copenhagen, Tokyo—great occasions, but they do not seem to have quite the historical appeal of those pre-war efforts.

Reading this near the end of the century, observers will surely note how little of what we would now recognise as medical physics appeared in the Congresses, although several reputable hospital physicists were there. At Zurich, it would look very much as if the presentation from Mayneord and myself was the only recognisable specimen. Physics there was, of course, but only marginally medical. The origins and properties of those mysterious mitogenetic rays were prominent, only to disappear; respect was paid to the biological and medical possibilities of X-ray crystallography. There was some physics under the heading of electrology and even heliotherapy, but these would not really be recognisable today.

More or less the same must be said of Chicago. There was some physics, even a little mitogenetic radiation, but recognisable medical physics was sorely lacking. This was partly due to the strong American influence at that Congress; in the United States, hospital physics, as we know it, was a very slow starter. The urgent dosimetry problems for

X-rays and gamma rays seemed to have been solved, but the general surge in radiological and other medical physics was still to come.

CHAPTER 11

NORTH AMERICA

Over the 60 or so years covered by these notes, the quite dramatic growth and development of medical physics as a profession is very obvious. In this country alone, from the half-dozen or so of us in 1932 there came in 1944 the 53 founder members of the HPA and now we have our own professional body, the Institute of Physics and Engineering in Medicine, with a membership in the thousands. Running parallel with the expansion of more sophisticated radiology, radiotherapy, nuclear medicine and medical engineering, physics showed a similar, though perhaps not always quite so dramatic, advance in many countries of the world. One consequence of all this was that from, say, 1950 onwards, when the noises of war had died down, there were tremendous demands for exchange of information in radiological physics as in other fields, between the developed nations. At the same time, the developing nations were clamouring for guidance and material assistance in these rapidly growing areas. One consequence of these developments was a constant movement of experienced scientists between developed countries, exchanging information, and to developing countries, assisting in the building up of their specialities. Some people recognised and found derisive amusement in the phenomenon of the travelling professor, or at his home base, the absent professor. Although only a small part of the scientific spectrum, this pattern appeared quite clearly in medical physics. Whilst still trying to avoid any formal medical physics history, a few comments and anecdotes concerning travel with medical physics implications might be permitted. For the sake of clarity, I propose to limit this chapter to North America.

In the field of the radiological sciences, our relations with North America were very close and I visited the United States and Canada no less than four times between 1952 and 1960. In these days, when

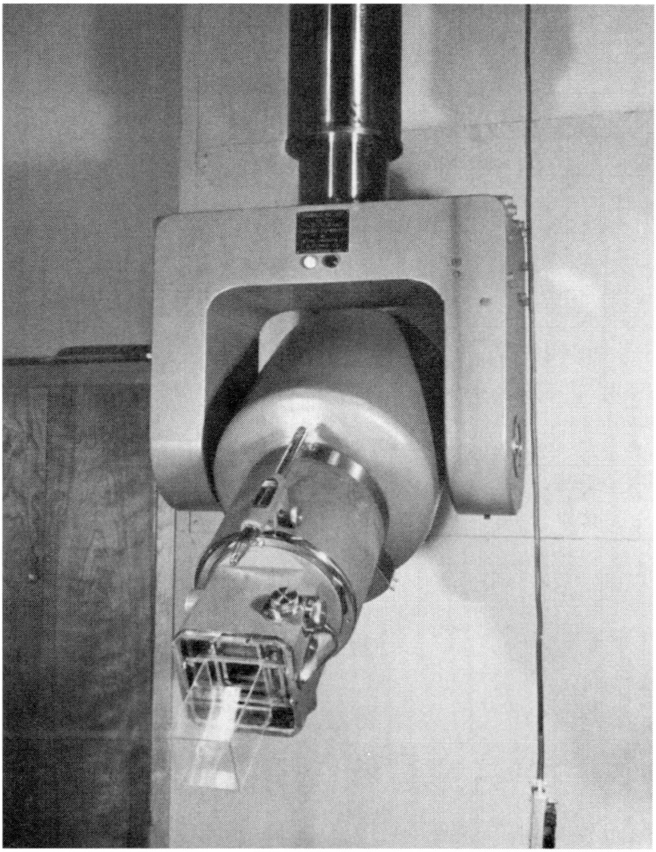

Canadian kilocurie cobalt units: Saskatoon, 1000 curie cobalt telecurie unit, reproduced from p 67 of 'New lamps for old' Radiography, April 1955.

a day return by Concorde is quite feasible if you can afford it, such travel might seem trivial, but not in the 1950s.

11.1 Gamma-ray-emitting Radioisotopes: Cobalt-60

The prospect of the routine use of gamma-ray-emitting radioisotopes for conventional beam radiotherapy was the reason for my first post-war transatlantic visit. We still had our 5 and 10 gram radium 'bombs' but therapists were dreaming of very high-activity gamma-ray sources which could be used in large 'telecurie units' (bombs), with outputs and focal distances comparable with those of conventional X-ray equipment. This meant kilocurie sources, not a prospect to cheer a physicist responsible for protection.

By about 1950, it was clear that the potential production of cobalt-60 could easily satisfy this demand. The main progress in that field had been made at Chalk River, with the nuclear reactors of the Canadian Atomic Energy Establishment. By 1952, the associated commercial company, Atomic Energy of Canada, Ltd (AECL), was producing the first commercial cobalt telecurie units. At that time, my colleague, Brian Windeyer, and Sir Stanford Cade were visiting Canada and returned with the news that a Canadian newspaper owner, Mr O'Connell, had offered to present one of the units to Britain. The unit was to be offered to the British Empire Cancer Campaign and the Campaign decided it should be installed at Mount Vernon Hospital in Northwood. In early 1953, the Campaign asked me to go to Canada to pick up at first hand technical information at Chalk River and at the AECL works in Toronto. I would also visit the two or three hospitals where cobalt units were already in clinical use. This was an opportunity not to be missed and on 7 April 1953 I embarked on my second North American journey, believe it or not in these jet-propelled days, sailing for Montreal from Liverpool on Cunard's *Samaria*.

The technical notes of this journey have long disappeared and become technical reports. The main idea of the visit was to ensure that when the new cobalt unit arrived in Britain, at least one person would have some idea as to what it looked like and what should be done with it. After some tribulation, I eventually reached Ottawa, where I received a most friendly welcome from the directorate and staff of AECL. Although nothing specific was said, I have no doubt that, to my hosts, I was the first glimpse of a potentially attractive profitable market for the telecurie unit in Europe. This, of course, did happen. As I might have expected, the new machine was a large and very sophisticated big brother to our old Bryant Symons 10 gram radium telecurie unit. From Ottawa I was whisked away for a few days in Chalk River, site of the AECL nuclear reactor and nuclear physics centre, way out in the outback (sorry, that is Australian) of Northern Ontario. This was an impressive technical experience though I have to admit that my clearest recollection of Chalk River is that there I saw and ate by far the largest steak I had seen for 10 years. Meat was still rationed in Britain.

Having seen the cobalt unit being designed and built in Ottawa and Chalk River, it was time to see it in hospital practice. First step on my way West was Toronto, though my recollections of the scientific aspects of my visit there are very limited. There was, however, much discussion about the planning of major new radiotherapy and medical physics departments. Even the subject of a chief physicist for the department was raised. The sky was the limit and a number of leading British hospital physicists would have been surprised to know that their names were bandied about. There was even a suggestion that I should resign my London Chair and emigrate. Eventually, the appointment which seemed to me the most reasonable and sensible was made and Harold Johns was moved from Saskatoon to build up an outstanding department. Perhaps the main reason for calling at Toronto was to make a side trip to London, Ontario, to see my first cobalt unit in clinical use. This was naturally an exciting experience, and I was surprised to find the process of setting up patients for treatment remarkably simple. My main recollection seems to have been the public reaction to the unit. Apparently the whole North American press had gone to town on this Canadian alleged miracle cancer cure. I was told of huge numbers of pathetic letters from all over Canada and the United States, appealing for priority treatment, some offering appropriate fees. With the best will in the world, even with the unit working 16 to 18 hours a day, the Victoria Hospital could only cope with a fraction of the demand. I was even told, although I could not confirm, that one of the leading hotels in London, Ontario, was offering special package deals for 'out of town cobalt patients'.

My day trip to London (Ontario) was followed by the long and incredibly beautiful train journey across the prairies and the Rocky Mountains. There is a great temptation to try and describe this but it is not medical physics and, in any case, far more able pens than mine have already done it. That long journey was broken for a few days for an important stopover in Saskatoon; here Harold Johns was in charge of medical physics so it had to be exciting. His scientific and technical contributions at Saskatoon and later at Toronto are well recognised and he is indeed one of the real characters of medical physics history. Writing this, my mind immediately flies to two quite unconnected Johns anecdotes. I think of the International Congress of Radiology in Copenhagen and of the late evening after the Congress dinner. Many of the Congress delegates had moved out into the famous Tivoli Gardens only to find Harold and myself, in full dinner dress, cavorting and colliding on the 'dodgem cars', to the delight of the populace. Incidentally, we were both stone-cold sober. On another occasion, Harold was visiting with us at our lovely old house at Osterley in West London. I can only think of the 'polite' anger and disgust of my young daughters when Harold persisted in referring to our beautiful old Victorian walled garden as our 'back yard'. I suspect he was never forgiven. At Saskatoon, he had jumped the gun on AECL and had built what was really the very first kilocurie cobalt therapy unit. In fact, this had the great disadvantage of being a fixed beam unit, with consequential setting-up problems. Nevertheless, I picked up much valuable information in Saskatoon on such things as dose measurement of these high-energy radiations, the planning of treatments and particularly the general safety precautions with such large machines. Our experience in Britain at that time was, of course, with much smaller units.

On then, over the prairies, followed by a spectacular train journey over the Rockies to Vancouver. Here another of the AECL cobalt units had been in clinical use for some time. I was very soon conscious of what I might call the politico-commercial aspect of the cobalt business. Vancouver was very conscious of the honour of pioneering this new cancer treatment and it had tremendous press publicity. It was clearly important to impress the visitor from Britain who would take the message across the Atlantic. The result was a very warm reception from everyone at the Hospital and in the city. I was persuaded to hold a formal press conference in my hotel room and this almost ended in disaster. For starters, I made an offer of drinks to the press and saw eyes light up. When I put my hand on the phone to order drinks, faces fell with a thud; at that time it was illegal in Canada to order drinks directly from an hotel room and it was clearly too late to send out for bottles. The press conference proceeded in a rather cool fashion thereafter with coffee, but the following day the press reports were quite impressive, with one exception. The photograph of myself reproduced in some papers was, to say the least, unflattering. Was it gentle malice or incompetence?

11.2 Renewing Old Contacts

The official part of my cobalt-inspired trip ended in Vancouver but I managed to move over into the United States and make another spectacular transcontinental journey in the *Hiawatha* from Seattle to Chicago. Here I was able to renew contacts with old acquaintances in radiotherapy and medical physics. The cobalt unit fever of Canada had not quite reached the United States. I suspect it never really did; electron accelerators were already approaching the battlefield. I believe Failla was toying with the idea of a super radium unit but it never appeared. The Francis Delafield Hospital had ordered a Canadian unit so that I was welcomed by Carl Braestrup, then physicist there, and we had much to talk about. My journey home was on the *Queen Mary*, my second trip on her but this time first class instead of steerage.

When the Canadian kilocurie cobalt unit was eventually delivered for installation at Mount Vernon Hospital, I was there and had the great advantage of being familiar with the new monster. The installation was relatively simple except that, for at least a minute, we held our breath. For some reason, the capsule containing the cobalt-60 stopped just as it left the transport container and before it entered the treatment head. For what seemed an age (probably a minute or two) our team was standing some three or four feet from a naked source of at least 1000 curies of cobalt. As it happens, I was, in fact, the nearest. The doses recorded on that occasion were never made public.

The 8th International Congress of Radiology in July 1956 assembled in Mexico City, perhaps a rather surprising venue, Mexico not being one of the growing points of radiology. For me, in my medical physics meanderings, it provided a fascinating experience not always connected with professional matters. Mexico was rather remote to justify asking permission to attend even so prestigious a conference. Quite unexpected relief came with an invitation to take part in a seminar at the Medical Division of the Institute of Nuclear Studies at Oak Ridge, Tennessee, an invitation which would make a very substantial contribution to the expenses of the Mexico trip. Although air transport was now common practice, I was again to cross the Atlantic by sea, this time in the pride of the American merchant fleet, the SS *United States*. Travelling excitement started at Waterloo Station, when going for the boat train for Southampton. The Station Master appeared in full regalia, complete with top hat. Then there arrived President Harry S Truman and his wife, who were to be fellow passengers. Yes, I had a word with the President during the voyage; medical physics was going up in the world.

Rereading my travel diary for the first part of this trip suggests a rather limited attention to medical physics. Arriving in the United States on 3 July was perhaps a mistake in itself, the following day being a national holiday for Independence Day. On the morning of the 4th, I travelled to Norwalk, in the heart of New England, and spent the day with some non-professional American friends. For most of the day we were driven around New England by my host, who had a special flair for arriving regularly at places where the British were allegedly humiliated during the War of Independence. In New York I seemed to devote a lot of effort in trying to find Professor Rotblat. Just as these notes are being written, it is a matter of great pride to record that Joseph Rotblat, a good friend and a great medical physicist, is being honoured with the 1996 Nobel Peace Prize. Another day was spent in New England at Yale, visiting biophysicist Professor Ernest Pollard, a very good friend from my long-past postgraduate days in Leeds. Incidentally, I caught up with Joe Rotblat there.

More interesting in a medical physics context was a visit with John Laughlin at the Sloane Kettering Institute. Here they were struggling with the problems of dose distribution and treatment planning associated with rotation therapy. The system was most complicated and elaborate, involving polar coordinates. The estimation of patient dose involved the use of thousands of punched data cards carrying dose distributions for all the axes at half-degree intervals under various conditions. The cards were then passed through a full-scale mechanical sorting machine and the outcoming data processed on a huge electric accounting (adding) machine and transferred to paper by hand to derive isodose curves for the patient's treatment. These sentences, copied almost verbatim from my 40 year old travel notes, must sound prehistoric to our modern radiological physicists with their bench-top computers and their customised planning software.

In New York, I entertained one of my post-war group of budding hospital physicists (Chapter 6), Ralph Worsnop, then in Philadelphia though now a planning software mogul in California, and his wife, my ex-secretary, to lunch at Jack Dempsey's restaurant on Broadway. I even shook hands with the great fighter.

After four days in and around New York, I began the long trail by rail to Oak Ridge, Tennessee, and then on to Mexico. With a brief relaxation weekend in Philadelphia, Washington was the next stop. A short stay only there but it was an opportunity to catch up with old acquaintances, like Harold Wykoff and Herb Attix, not strictly medical physicists but major contributors to radiation dosimetry, and particularly Lauriston Taylor, who predated me in the dosimetry business. It was also a chance to see for the first time that wonderful complex of hospitals and laboratories which form the National Institutes of Health.

The next step in this mixed-up journey to Mexico was Oak Ridge, which, at that time, we tended to think of as the American equivalent of Harwell. There, Marshall Brucer, one of the undoubted characters of radiotherapy, and Director of the Medical Division of the Oak Ridge Laboratories, had realised that large numbers of radiotherapists and their physicists from Europe and North America would be passing through and over Oak Ridge on the way to the Radiology Congress in Mexico City. The Atomic Energy Commission had agreed to subsidise a considerable number of these travellers for a stopover in Oak Ridge, with a seminar for a preliminary discussion of Congress problems. Thus, on my first visit to the old Alexander Hotel in Oak Ridge, with its Wild West appearance and its rocking chairs on the veranda, I was soon joined by familiar faces: Tom Chalmers of Liverpool, Frank Farmer of Newcastle, Frank Ellis, then of Oxford, David Smithers, and there were others. There was also the opportunity to meet a number of American physicists who had previously been just names in journals.

11.3 Roentgens, Rads and Riddles

At that time, 'supervoltage' was fast becoming the 'in' word in radiotherapy; cobalt units, linear accelerators and a few betatrons were heading for the situation where they were to replace the 200 kV machines as the workhorses of radiotherapy departments. Rotation therapy had been introduced and the dosimetry thereof was giving many headaches to the physicists involved. Indeed, I read a short contribution on this subject myself. I fear this might be considered rather elementary to the computer-rigged physicist of the 1990s. Nevertheless, a major part, though not quite the whole, of our Oak Ridge seminar covered these subjects. We did not solve many of the problems but the results of the discussions appeared in an oddly (Brucer-like) named volume published by the Atomic Energy Commission, the title being *Roentgens, Rads and Riddles*. Maybe a few copies survive in some radiological archives; it is quite a good read. It was at this seminar that Brucer, during one of his lectures, produced a slide entitled 'L'assassin du Röntgen', which created a great deal of surprise and amusement. The slide showed an obviously barred prison cell window, with a prisoner staring out from behind the bars. To my own amazement and to the hilarious delight of the audience, the prisoner behind the bars was none other than myself. I have never been convinced of any reason why I should deserve that stigma, nor where Brucer found that superimposed photograph.

11.4 Mexico Congress

After the delights of Oak Ridge came the long and, at times, rather dreary journey by rail to Mexico City, fortunately interrupted by a delightful day looking around Memphis. My first impression of Mexico City was of the breathlessness arising from some 7000 feet altitude. I had been warned of this and also of the horrors of Montezuma's Revenge (aka Gippy Tummy in other parts of the world). Against the latter, I had taken adequate precautions but more than half the Congress members suffered. I imagine that, from a radiological point of view, Mexico 1956 would be regarded as a good, but not outstanding, Congress. There was quite a large medical physics presence, perhaps a little surprising in view of the distances involved. Their contributions were, to say the least, mixed. On one day I listened to some four physics lectures and made a single comment on each, 'not original'. I shall refrain from mentioning names. Writing this some 40 years later, I was impressed by a note on another group of lectures. Brownell talked of neutron capture therapy, Rotblat outlined the problems of a 15 MeV accelerator and K C Tsien (then at Sloane Kettering, later at the International Atomic Energy Agency) spoke on the possibilities of computers in radiotherapy dosimetry planning. From the vantage point of the 1990s, we can see that radiological physics was on the threshold of major technical advances, of which we can see the results today.

As might be expected in Mexico City, the social additions to the Congress were not neglected. These included an exciting and colourful Mexican rodeo, a very long tour of the University, the almost incredible shrine of Guadeloupe and the Mexican pyramids.

The journey home from Mexico City was surely something to be remembered. The whole trip as far as New York was done by train, with no stopover, and took three days and three nights, occasionally wearisome but very comfortable. Then home again by sea in the MS *Flandre*, just a bit of French luxury.

11.5 Iodine-131

Within a year I had meandered back to Oak Ridge for yet another seminar. The agenda for this meeting was much more limited than that of the previous year. What we now call nuclear medicine was steadily advancing but by far the largest section of the subject was still concerned with iodine and thyroid metabolism and pathology. The *in vivo* measurement of radio-iodine had been developed in many major departments of medical physics around the world and there was some feeling that some standardisation and intercomparison was desirable. To discuss this and allied problems, representatives from a wide range of departments were invited to Oak Ridge with the generous assistance of the Rockefeller Foundation. I must admit that I had some doubts about my own qualifications for such an invitation. At that time, the Oak Ridge Medical Division was working on the iodine-131-loaded 'human manikin', which some older colleagues would remember was circulated to hospital physics laboratories around the world for cross checking and standardisation of radio-iodine uptake measurements.

11.6 *Physics in Medicine and Biology* and Robert Maxwell

This particular transatlantic trip was not limited to Oak Ridge. Courtesy of the Rockefeller Foundation, I was able to meander round quite a number of hospital physics departments. Reading the notes of that tour reminds me that the mid 1950s were a most important period in the development of medical physics as an independent profession. Following the formation of the HPA, new national bodies arose and

CONTENTS

Page

Editorial 1

Recent Advances in Electrophoretic Separation Methods for Biologically Important Substances. By C. J. O. R. Morris, Ph.D., F.R.I.C., Department of Experimental Biochemistry, The London Hospital Medical College, Turner Street, London, E.1 3

The Present Status of Calibrations of Ionization Dosemeters with High-Voltage X-Rays. By G. P. Barnard, D.Sc., Ph.D., M.I.E.E., F.Inst.P., E. J. Axton, M.Sc., A.Inst.P., D. S. C. Belcher and A. R. S. Marsh, The National Physical Laboratory, Teddington 18

Microspectrophotometry with the Burch Reflecting Microscope. By K. W. Keohane and W. K. Metcalf, Department of Anatomy, University of Bristol 25

A Profile Counter and its Calibration. By B. D. Corbett, B.A., R. M. Cunningham, M.D., D.M.R.(T.), K. E. Halnan, M.B., D.M.R.(T.) and E. E. Pochin, M.D., F.R.C.P., Department of Clinical Research, University College Hospital Medical School, London, W.C.1 37

Analysis of the Radioactive Content of Tissues by α-Track Autoradiography. By J. Rotblat, Ph.D., D.Sc., F.Inst.P. and Gillian Ward, M.Sc., Ph.D., Physics Department, The Medical College of St. Bartholomew's Hospital, London 57

Instrumental Notes. Some Recent Nucleonic Instruments for use in Clinical Work. By E. W. Pulsford, B.Sc., A.M.Brit.I.R.E., Atomic Energy Research Establishment, Harwell 71

Reviews of Books 74

Abstracts Section 75

Table of contents from the first issue of Physics in Medicine and Biology, Volume 1, No 1, July 1956.

proceeded to internationalise the scene. My tour of the eastern states had an adequate scientific and technical content but my diary notes contain constant references to discussions of the professional bodies and the possibilities of cooperation. I had no mandate to discuss these matters officially but, writing some 40 years later, it is good to record the tendency. Furthermore my beloved *Physics in Medicine and Biology* was first published in 1956 with myself as Founding Editor, and this had clearly aroused much interest. As soon as I arrived in New York, I was invited to Robert Maxwell's Academic Press for discussions about PMB. I can now find no record of these discussions but PMB firmly stayed with the HPA and its then publishers, Taylor & Francis. It is naturally a great joy to me, having suffered the tribulations of those early years, striving to keep the journal alive, to see the present-day PMB, a highly respected and truly international journal, covering the whole world of medical physics.

My return journey from Oak Ridge allowed a stopover of several days in and around Pennsylvania, where I could find much of interest. I was able to make contact with Dr Herman Schwann of the University of Pennsylvania School of Medicine, who had been of great assistance to us at the Middlesex in the building up of Edward Grant's work on the dielectric properties of biological materials. Another unexpected renewed contact, social rather than scientific, was with one of the real pioneers of medical physics, Edith Quimby of New York. There is even a note that I lectured to the radiologists of the University of Pennsylvania School of Medicine, but I cannot find any record of the subject of such a lecture. One day was given over to a visit to Baltimore and the Johns Hopkins Hospital. Even at this temple of academic medical science, there was much talk of the HPA, The Biophysics Society, The Health Physics Society and others, with kindly references to PMB, and appropriate hopes of collaboration.

This tour ended, in fact, with great excitement when the *Queen Mary* was kept out of Southampton by a dock strike and eventually a small Cunarder had to ferry us from Cherbourg to Plymouth.

11.7 1959

My next visit to the United States in 1959 was quite different again and perhaps justifies a paragraph or so. I had another opportunity to meander across the United States and take up some invitations to visit and even lecture. The scientific content of the trip was rather limited, but it provided a surprising number of irrelevant anecdotes for future reference.

Dr Edith Quimby, reproduced with permission from the HPA Bulletin, March 1983.

As recorded elsewhere in these notes, I was invited for the second time to travel to Hong Kong to act as physics examiner for the Diploma in Medical Radiology students who had studied in the Far East. By that time I had come to terms with the idea of flying on long-distance visits and, when making plans, it again seemed rather ridiculous to sit in an aeroplane and fly straight home after the examination. Much better to come home the wrong way round and make some of those promised visits in the United States. Thus, examinations over, Japan Airlines picked me up in Hong Kong to take me to California, and it turned out to be quite a journey. For instance, on a long stopover in Tokyo, I learned that the Japanese drive their cars on the left of the road, just like us!

In those days, aeroplanes had propellers and a limited range; this gave me the exciting experience of landing on that tiny dot in the middle of the Pacific Ocean, Wake Island. Hawaii almost brought personal disaster. I had obviously eaten unwisely in the Hong Kong floating restaurant and my stomach was in the process of revolt. It was clear as we approached Hawaii that I could go no further. Japan Airlines sprung into action: doctor at the airport, taxi to an hotel, room booked for two nights and onward flight planned. This is the story of my first and only visit to Honolulu and Waikiki beach. Even the journey on to California was unusual: no first-class seat was available so I was allocated a bank of three standard seats so that I could lie down.

In California I reverted to my normal state as a medical physicist, gazed in wonder in Palo Alto at my first underground particle accelerator, where particle path lengths were measured in furlongs. I responded by giving invited, though not very exciting, lectures in Los Angeles and San Francisco.

Oak Ridge had to be included in my meanderings and somehow I was booked on a very domestic flight to Knoxville, a 'milk run' hop along the Mexican border in an almost empty plane, landing every half hour or so. We even lost an engine at one point and had to wait for a spare plane. Then from Nashville, the plane was crammed with a crowd of noisy young men. It was soon obvious that they were on their way to a football game in Knoxville.

I must admit that in 1959, travelling by air to a local football game was way beyond my sporting experience. More exciting, perhaps, was the transatlantic journey home. Older readers may remember that about that time, one of the airlines in the United States was very publicly planning and advertising the introduction of the very first jet-powered service on the Atlantic run.

Apparently, under strict secrecy, BOAC were modifying and preparing some of their Comet fleet for just such a service. Then, without advertising or warning, they introduced a daily jet service well ahead of the Americans. The reason for introducing this irrelevant item here is that I was offered and accepted one of those early flights and it was, indeed, a most exciting experience.

11.8 Conclusions

A postscript to these journeyings around North America would seem to be desirable. It may well be argued that these particular descriptions of my travels in that part of the world contain too little of what we now regard as medical physics. I have to admit the truth of this point and the simple explanation is that it was just not there. As a separate and recognisable profession, medical physics in Britain expanded very rapidly from the time of the formation of the HPA with its 53 members (not all strictly hospital physicists). This kind of development came much later in some other parts of the world and particularly in the United States. In Britain, with its social medical services and, a little later, the National Health Service, there was no serious problem. The hospitals could very simply employ physicists to join the clinical teams providing radiotherapy and later other services. In the United States, the dominance of private medicine made this much more difficult. Who was to pay the physicists? Much of radiotherapy was in private hands and there were obvious difficulties in employing physicists. In the text of this chapter, I report that Milton Friedman tried to persuade Herbert Parker and myself to move to the United States but there was no clear indication as to how we should be employed. The problem is also highlighted by the experience of one of my 1950s' staff, Ralph Worsnop. He was a competent and well-trained hospital physicist when he emigrated to the United States. Instead of finishing up in a hospital post, he set up a prosperous business in California, producing and selling computer software for use in radiation treatments.

This does not mean that no medical physics development was in hand. At this time in the United States, betatrons, linear accelerators, electrostatic generators and so on were being developed but all this was happening in universities and research institutes with, I suspect, little thought for medical development. With the possible exception of the electrostatic generators in Boston, there was, indeed, little or no medical connection. Hospital-appointed physicists were almost non-existent. It could be said, though, that medical physics in Britain received one of its main boosts from these high-energy developments. Wider horizons outside radiotherapy followed almost automatically.

CHAPTER 12

FAR AND WIDE

12.1 The Curie Institute

I outlined my first experiences of overseas meanderings in Chapter 10 with my attendances at the International Congresses of Radiology in Zurich (1934) and Chicago (1937). There was only one more foreign adventure before the war and this must be recorded from memory as all the paperwork has disappeared. Some time around 1938, I received a most surprising invitation to spend a few days at the Curie Institute in Paris, to discuss with the staff there the acute problems of the dosimetry of radium gamma rays. To suggest that the invitation was a complete surprise was indeed a gross understatement; for me to be asked to go to the fountain-head of radium studies and talk about radium dosimetry seemed quite beyond my imagination. In the end, I suppose vanity overcame modesty and the invitation was accepted. Here I made a very serious mistake indeed. The letter of invitation was, of course, written in French and, in a further fit of vanity, I asked the Department secretary, who spoke fluent French, to translate my reply before typing it. The end result can be imagined. I duly took the through train from Waterloo to Paris (no Channel Tunnel but trains then crossed the Channel uninterrupted, by ferry), and presented myself at the Curie Institute. My welcome there was tremendous and most friendly, from the girl on reception to the Directeur himself, the great Antoine Lacassagne, everyone naturally speaking in fluent colloquial French. In modern terms, my own French may have been described as rather poor GCSE level. The technical part of the visit was quite a success, with a series of meetings and discussions with the Institute staff, all convinced I was fluent in French. In fact, I managed quite well with the cooperation of a Spanish visitor who spoke some English and German. I tell only one clearly remembered story which older radiologists who knew Antoine will appreciate. I stayed for the weekend in Lacassagne's apartment in Montparnasse, a delightful experience in itself. When I was leaving, he called a taxi to take me to the Gare du Nord and I quite clearly heard him whispering to the driver that I was running late. This was not entirely true but it was Lacassagne at work. Anyone who has driven in Paris will appreciate the reactions of a French taxi driver and the subsequent 20 minutes of terror which followed.

12.2 Development of Medical Physics Overseas

In the interests of coherence and to avoid clashes of time and place, in Chapter 11, I limited my descriptions of post-war meanderings to North America. During the war and immediate post-war years, the development of medical physics, although not generally hospital based, followed much the same routes in North America as in Britain. Post-war visiting between the two areas, as Chapter 11 shows, was largely a matter of conferences, lectures and seminars, aimed at exchange of information. In other parts of the world, and particularly in the developing countries, medical physics was part of the general demand for scientific and technical updating and assistance. The phenomenon of the 'absent professor' travelling the world on 'aid' missions was very obvious in our field. It should be realised that the development of what we now understand as medical physics was very limited in many places outside Britain and this puts serious limitations on these notes.

It happens that the first long journeying which comes to mind had little or nothing to do with scientific or technical development. It was a largely academic radiological affair. At that time in the 1950s, the postgraduate Diplomas for Radiology, the DMRD and DMRT, were still in the hands of the Royal Colleges of Medicine and Surgery, through the Faculty of Radiology, before they were eventually handed over to the Royal College of Radiologists. A substantial number of candidates followed their training courses in the Far East and then came to London for examinations, a most uneconomical procedure. In 1958, the Colleges decided that the time had come to hold an examination for Part 1 of the DMRD and DMRT in the Far East and chose Hong Kong for this purpose. Thus Dr Rohan Williams of St Mary's, Paddington, and I were invited to go to Hong Kong to supervise the Part 1 examinations and act as external examiners. This was an opportunity not to be missed and, for me, it brought about one major

Presentation of an ivory statuette to Professor J E Roberts and Dr E Rohan Williams by the Hong Kong branch of the British Institute of Radiology.

change: I dropped my objections to air travel. The journey to Hong Kong by sea or land took several weeks and the choice was simple—fly or lose out.

12.3 Hong Kong

The start of that trip was not auspicious. A week before departure there was an airline strike; the whole of BOAC's Britannia fleet was grounded and our trip was off. Then, at the last moment, we were phoned and asked to assemble at Victoria, were loaded onto coaches and driven past the lined-up Britannias at Heathrow to Blackbush Airport where we were all loaded onto an old chartered Hermes. Taking off from Blackbush in a noisy Hermes was, for me, something of a terrifying introduction to air travel, but we arrived in Zurich to find a stray Britannia to take us on for the further two days plus to Hong Kong. This paragraph may not be medical physics but it was certainly meandering on the grand scale.

A first visit to Hong Kong is an unforgettable experience and description is quite beyond my pen. We had to work very hard at invigilating the written examinations, preparing and supervising practical tests, marking papers and oral examinations. All this work was interspersed and well compensated by a fascinating round of visits and entertainment. We were housed at the luxurious Repulse Bay Hotel, a few miles out of the city on a delightful beach. There was an almost fearsome round of lunch and dinner parties where I learned, temporarily, to use chopsticks. The Hong Kong government, which was responsible for our hospitality, really did us proud. Even the Governor summoned us to lunch at the Mansion on Hong Kong Peak. Medical physics was going up in the world.

There was a great finale to the entertaining of the examiners. All the DMR candidates, successes and failures, invited the examiners to dinner at one of the famous Aberdeen Hong Kong floating restaurants. Indeed I have a picture of the occasion showing myself consuming a giant prawn with the aid of chopsticks—surely a time to be remembered, even long afterwards. In front of me as I write is a beautifully carved traditional Chinese ivory statuette, one of which was presented to each of Rohan Williams and myself by the Hong Kong Branch of the BIR. Obviously they really appreciated an unprecedented visit from not one but two ex-Presidents of the BIR at the same time.

This was, of course, not strictly a working medical physics journey and so far I have not even mentioned the subject in connection with Hong Kong. Nevertheless the subject was very much alive, largely due to the presence of one person, Geoff Maulden, an expatriate Australian physicist who devoted the greater part of his professional career to physics in Hong Kong hospitals. A Chinese-speaking Australian in Hong Kong was quite a character to know.

During the 1950s a number of Hong Kong physics graduates had received basic training in radiological physics at the Middlesex and elsewhere in Britain and there was great pleasure in renewing contacts. At that time a large new hospital complex was being planned for Kowloon, with a major new radiotherapy department. Even then there were two linear accelerators on the shopping list. Hong Kong may have made no great original contributions to the medical physics of the time but there was always a very adequate hospital service, including a developing department of nuclear medicine.

Before planning the Hong Kong visit, the thought occurred that simply to return home after the examination seemed rather a waste of the considerable travel budget. I had ideas that a little meandering around Australia to see the development of medical physics, together with a brief holiday, might be a good thing. The Medical School Council readily agreed to extra leave for this purpose.

12.4 Australia

So to Sydney it was, with just three or four days to study how medical physics might be progressing in far-away places. Writing 40 years later, it would be quite wrong to comment on individual works. It would seem that in the previous few years Sydney had imported several English academic physicists of some standing into the hospital service. This probably accounted for the somewhat greater emphasis on research than one would expect in a hospital at home. Rightly or wrongly at the time, I got the impression that conventional hospital physics, closely related to radiotherapy as we then knew it, was regarded by many as a second-class option. On the other hand, I found one physicist, who shall be nameless, prospering mightily on iodine-131. Apparently he had a very lucrative practice, partly hospital and partly private, cooperating with the physicians on iodine uptake, distribution and so on. Apart from the signs of technical growth, socially, Sydney provided some happy meandering.

The next step was a short hop to Melbourne for a similar brief study. Even just going to Melbourne at that moment presented problems. Unknowingly, I had chosen to go in the week of the Melbourne Cup (Derby week to a simple Englander), and that meant inevitable chaos. I was fortunate in being able to get a modest hotel room but unfortunately had to report the loss of my wallet from that same locked hotel room. Sorry, Melbourne.

The medical physics situation was encouraging. I was greeted on arrival by Ken Clarke, one of that remarkable group of young Middlesex physicists mentioned earlier, who had gone out to Australia to join Jack Martin at the Melbourne Cancer Institute. This was one of the largest radiotherapy departments in the world, with nine 200 kV conventional X-ray machines and a linear accelerator. It was clear that, unlike Sydney, medical physics in Melbourne, as in Britain, was, for the time being, to be closely linked to clinical work and particularly to radiotherapy. Nuclear medicine was soon to follow.

I took time off from the hospital activities and accepted invitations to Melbourne University to see some 'real' physics with an electron synchrotron and a 36 inch cyclotron, fascinating work on proton scattering, photodisintegration, cosmic-ray plates in balloons and much more. The whole University set-up clearly made a great impression. In my travel notes I find here and there phrases such as 'would not mind coming out here', and even at one point a note about the possibility of asking for a sabbatical to be spent in Australia. Perhaps some of these thoughts were inspired by a five minute visit to the Melbourne Cricket Ground, walking the hallowed turf as it was being prepared for a test match against England. Eventually I got my sabbatical but I ended up in Beirut (Chapter 13). I ended my visit to Melbourne by giving a lecture to the Radiation Society on the then recent Adrian Report on Radiation Doses to Populations. With hindsight, this sounds a little presumptuous as, although my Department collaborated and provided staff and facilities for the Adrian team, I was not closely or personally involved in the study.

My last official stop on this Australian trip was in Adelaide, where I could forget hotels and lost wallets and stay with relatives. This leads to one small piece of name dropping which I cannot resist. In Adelaide, I was taken by my family host to the local Rotary Club lunch. Here I was introduced to and sat and enjoyed my lunch with the great Don Bradman. There was much to be said about cricket and particularly Yorkshire cricket. Here was a story to be told to my great-grandchildren.

From the point of view of radiotherapy and its associated physics, Adelaide was on a much smaller scale, though with 250 kV X-ray machines and also a linear accelerator and cobalt unit on order. The small physics staff was able to produce some valuable work on ion chamber calibration. A visit to the Radiotherapy Department Common Room for an Australian-type cup of tea brought surprises. Three or four of the staff had been through our Department at the Middlesex and there was a great exchange of news and names. At one stage I was invited to a clinical staff meeting to discuss new patients, a rather rare event in Britain at that time. My diary sadly relates 'contributed nothing to the discussion'. A final visit was to the University Physics Department on the campus adjoining the Royal Adelaide Hospital. There I was greatly impressed by the extensive work in biophysics, well outside my personal comprehension.

A small postscript may be of interest. During the evening of my last day in Adelaide, a reporter from the *Adelaide Advertiser* called at my home, looking for a story about my visit. Clearly, my host, a well-known Adelaide business man, had been talking. I did my best to oblige the young reporter but eventually he defeated me with the question 'did I know anything about a report of the use of "sound echoing" instead of X-rays in obstetric imaging?' I had to admit that I did not. However, this was 1958 and only recently I discovered that in that year the seminal paper by Donald, McVicar and Brown appeared in the *Lancet* and for many years Sydney led the way in ultrasound research. The young reporter was clearly on the ball. In fact, the *Advertiser* published a very good paragraph and quoted me remarkably accurately. This particular trip brought to an end my long-distance voyaging by sea. I returned to England by the fastest possible mail steamer of the Orient Line in a little over three weeks. How things have changed.

12.5 Hong Kong Revisited

I was invited to repeat the visit to Hong Kong for DMR examinations the following year, but to try to describe this trip in detail would, I fear, be far too repetitious. I must confess that my clearest recollection of this trip is of the most perfect meal of my life: purely Chinese at the home of Dr and Mrs H C Ho. Once again I felt that to fly out to Hong Kong, conduct an examination and fly straight home was an unprofitable proposition. This time I had a number of invitations for visits and lectures in the United States, so this seemed an opportunity not to be missed. The crossing of the Pacific to Los Angeles turned into quite an adventure and is described in full in Chapter 11, together with the tour from there to New York.

12.6 South Africa

One other major journey was concerned with the examinations for the Radiological Diplomas. There was a feeling amongst radiologists in

South Africa and particularly in Johannesburg that they should, like Hong Kong, have independent examinations acceptable to the Royal Colleges. The Board of the Colleges approved the idea in principle and suggested that I should go to Johannesburg to report on the first Part 1 examination and act as an external examiner. This gave me an opportunity to see a typical example of a local growth of medical physics. There was a very active department of radiotherapy, with a forward-looking team of radiotherapists, and this immediately generated a demand for a quality physics unit. Such a team grew up under Phil de Valence and the journals of the time record useful work in radiotherapy and dosimetry from that source.

I have one rather unhappy memory of that particular visit. The department had one or more cobalt units but in the very unhappy circumstances of the time and place, their use was strictly limited to white patients only; no treatment of coloured patients was permitted. My unhappiness was somewhat mitigated and my faith in humanity restored when I learned that one radiotherapist, who, of course, I must not name, but for whom I had great professional respect, insisted on treating a small number of coloured patients in secret in the hours of the night. It is good to realise that this is now ancient history. On a happier note, visiting Johannesburg enabled me to spend a long weekend in the Kruger National Park, comfortable in a very pseudo 'roundel' native hut—with full en suite facilities.

12.7 Malaysia, India and Singapore

In 1966 there came an opportunity to meander into a very different part of the world to observe and take part in the early development of medical physics in new and unexpected surroundings. Our British government Department of Overseas Development had received a request from the government of Malaysia for assistance with the starting-up problems of a large new department of radiotherapy and nuclear medicine. This was to be a piece of so-called Colombo Plan assistance and I was asked to take on this task. Thus, on 26 November I set out for Kuala Lumpur, not really knowing whither I was bound nor what I was really going to do.

Before relating this adventure, I must interpolate a minor piece of side-tracking. During the week before my planned arrival in Malaysia there was to be held in New Delhi an International Conference on Medical Education. As New Delhi is on a direct line to Malaysia, I was asked to join three professorial colleagues to form the Middlesex delegation to the Conference. The Conference was most impressive, with names which resounded in an international context. It was firmly impressed on me that there was much more to medicine than radiotherapy and medical physics. One day I deserted the Conference and accepted an invitation to visit Agra. Of course I was duly dazzled by the Taj Mahal but also had another pleasant surprise. In Agra there was a very healthy department of radiotherapy, in the care of Dr Haldar, a Fellow of the old Faculty of Radiologists and one-time DMRT student at the Middlesex. It is very tempting indeed to write that 'consequently', he had a competent physicist on his establishment and even loaned the said physicist as our expert guide to the Taj Mahal. This was clearly a very satisfactory day's wanderings.

After this most interesting side trip, the real work had to start. A warm welcome in Kuala Lumpur was assured. At least two of the radiotherapists, including the Director, Dr S K Dharmalingen, were the products of the Meyerstein Institute and the Senior Physicist, Lee Kean Chye, had spent two years in my Department. The position on the ground at the Kuala Lumpur General Hospital was that the building of a very fine new radiotherapy department, with a nuclear medicine annexe and incorporating appropriate physics facilities, was almost completed, I believe at government expense. It was soon clear that, apart from some desirable internal rearrangements in nuclear medicine, a very impressive layout had been designed. The real surprise came when I discovered that the main machines for radiotherapy had been on order for a long time and were already under construction in Germany and England. When I also realised that the machines in question were a betatron for X-rays up to 25 MeV together with two 6 MeV linear accelerators, surprise was a gross understatement. Furthermore, there was no one on the staff, clinical or in physics or engineering, who had any experience whatever of megavoltage machines or therapy. The position in nuclear medicine was a little better. This was in the capable hands of Dr Yi Sung Tack of Korea, but there was scope for a major update of equipment. Unfortunately, Dr Yi Sung died not long after this visit. At first contact, the whole business seemed a hopeless proposition or, at the least, a very severe challenge.

On that visit I spent some six weeks in Kuala Lumpur, partly with the architects and construction workers but mainly with the existing departmental staff, trying to define the immediate and future needs of staff and equipment. The switch from a conventional radiotherapy unit with about three orthovoltage machines to a very large set-up with state-of-the-art megavoltage equipment was clearly beyond local resources and would undoubtedly require considerable temporary international assistance. Not surprisingly, it was hard work trying to convince the people on the spot of the real extent of this need. They were very proud of their achievement so far. Everyday supporting equipment for radiotherapy, physics and nuclear medicine on the scale proposed resulted in enormous lists of requirements, such as a whole range of dosemeters and counters, workshop equipment and very much more. By far the greater part of my formal report to the Malaysian government at the end of this visit consisted of long lists of staff and equipment which were required before the new department, with its massive machines, could even hope to start operations. The best that I can say is that my hopes and suggestions were a little optimistic, though realistic in absolute terms

I returned to Kuala Lumpur some 18 months later in July 1968. The idea was that all the treatment machines, most of the service and measuring equipment and at least the basic staff would be in position and everything ready for commissioning. It was immediately clear that I had been far too naïve and had failed to remember that administrative machinery moves distinctly more slowly in the East than in Western Europe. I should have known better. The big betatron was already installed and handed over, the engineers having gone home to Germany. Indeed I was shaken to find that a small number of patients had already been treated on this machine. Apparently the engineers had left behind some output figures from their testing procedures and from these the staff on the spot had devised a few simple treatments. Not surprisingly, they achieved some public acclaim. Unfortunately it was necessary to bring an engineer back from Germany for a short time during my visit. The linear accelerator engineers had effectively finished and it was possible to carry out handing-over checks almost immediately.

At first sight this looked quite promising but there were complications. My earlier recommendations that there should be a considerable increase in professional clinical and technical staff, with more training, had not been implemented. Thus the staff adequate for a department with two or three orthovoltage machines was loaded with two extra linacs and a betatron. My most urgent original staff recommendation was that an engineer should be appointed and spend some months with the manufacturers of the machines and assist in their installation. A very good young engineer was indeed appointed but he took up the post on the very day I arrived for my second visit. So far as I know, he had never seen a betatron or a linac in his life.

We decided eventually that the only thing to do was that the two young physicists and myself should calibrate and commission the linacs as far as was possible so that they would be ready for limited clinical use. This meant that I should doff my advisory role and settle down to become a conventional hard-working radiotherapy physicist of the period. I admit this raised some eyebrows amongst the staff and I may have lost myself some professional standing. I can offer one small anecdote on this theme. One morning we found that the main fuses on the line serving the linacs had blown. I asked for fuse wire only to be told that the Hospital Engineer was the only person who could possibly mend a main fuse and he was away for the day. The only thing to do was to close down the work for 24 hours. This I could not swallow and I eventually found some wire and proceeded to replace the old-style fuses. It worked but there was consternation when the machines came back to life. The very idea that the Professor should or could mend a fuse was almost incomprehensible. There followed two or three weeks of really hard graft. Sufficient of the physics equipment which I had recommended was available and with the local physicists and a couple of radiographers we managed a full calibration of the linacs and produced a limited range of depth dose and isodose curves. Thus the very limited staff would be able to make a start on some supervoltage therapy and prepare for a full-range future. The story in the Sub-Department of Nuclear Medicine was not so happy. Much of the proposed new equipment had not yet arrived and the untimely death of Dr Yi Sung had brought work almost to a standstill.

I could not be too happy about my two excursions to Malaysia but there was clear hope for the future. Some two years later on my way home from the International Congress of Radiology in Tokyo, I managed to spend a few days in Kuala Lumpur and was encouraged to find that a real department was beginning to emerge. Some time later, a colleague in London told me quite unofficially that he had been in Kuala Lumpur and had been impressed by the range and quality of the department's work. Perhaps to cheer me up, he told me, just slightly tongue in cheek, that, from time to time, the 'Roberts Report' of 1968 was brought out like a holy ordinance when further plans were being discussed. I was duly flattered. I must admit that I have written these paragraphs about Malaysia with some hesitation. My fear was that they might be interpreted as criticism of people on the spot in Kuala Lumpur. This is very far from the truth. Of course the plan for the new department was too ambitious but ambition with the best will in the world. It was good to know the many people involved, quite a number with Middlesex training and background, and I was proud to be attached to the programme, even though at times I stamped my feet and blew my top. It was all a brave attempt to spread the best of radiotherapy and nuclear medicine, with their attached physics, into yet another part of the world.

Singapore is, of course, only a half-hour hop from Kuala Lumpur and naturally I took the opportunity for visits. This was mainly to meet up with old friends but I also managed a little meandering in the medical physics field. Compared with Kuala Lumpur, this was on a very modest scale, with radiotherapy and physics developing together, very much in British style, advice and help also coming from the International Atomic Energy Agency (IAEA). I must admit that I found a distinct Middlesex atmosphere and I felt very much at home in Singapore, both in and out of the Hospital.

12.8 Egypt, Libya, Sudan and Syria

There is a region of the world where the development of medical physics and its associated clinical fields was and is slow and uneven. This might be oversimplified and referred to as the Arab world but I would prefer the more geographical Eastern Mediterranean and Middle East. Much of my own involvement in this area is covered in Chapter 13, with details of a sabbatical year (1963–1964) in that part of the world. My contacts and meanderings there, however, started much earlier, particularly in Egypt and Sudan. The links began in the early 1950s when Kemal Mahmoud (later Professor in Cairo) and Gamal Osman, sponsored by the Egyptian Atomic Energy Authority, joined my group of young graduates as trainees and later as postgraduate students. In

1956 I received an invitation to take up a short-term appointment as Visiting Professor of Physics in the University of Cairo. The main idea was to introduce some radiation physics into the undergraduate curriculum and to initiate a potential nucleus of graduate physicists for the benefit of the Egyptian hospitals which had radiotherapy. This was again a different approach to medical physics, preparing the staff before the equipment was installed. Other parts of the Arab world were expected to benefit, though I fear this ambition was never quite fulfilled.

As at that time I was still hesitant about flying, Egypt posed a problem—a few hours by air, a week by sea. For this occasion I found a compromise: cross the Channel by sea, Blue Train and Rome Express and on to Naples; overnight there and then catch the Orient liner, which had left London five days earlier, for Port Said and Cairo. It worked very well and my welcome to Egypt really shook me, being of high VIP standard. There was a delegation from the Presidential Council of Ministers who, to the amazement of my fellow passengers, took me off to be welcomed by the Governor of Port Said. I was then offered the use of the gubernatorial limousine, with escort, for a morning tour of the city. Medical physics was again going up in the world.

Arriving in Cairo for the first time must always be a moving experience, enhanced for me by a number of familiar faces, arising out of Middlesex contacts. I was soon allocated a room in the Physics Department of Cairo University in Giza; on the door was fixed a small piece of plywood on which was rather crudely painted my name in English and Arabic. When I eventually left, I borrowed a screwdriver and stole the plate; it still hangs over my desk at home today. One further slightly irrelevant point: I knew exactly where I was because if I looked out of my study window, I could just see the top of the great Cheops pyramid in the distance. Acceptance into the Faculty was rapid and I had to contemplate a lecture programme bordering on the horrendous, comprising undergraduates, postgraduates and staff in Cairo, a series of lectures in Alexandria and a scattering of public performances. The Egyptian Atomic Energy Authority, with its reactor centre at Enshass and laboratories in Cairo, was largely responsible for all uses of ionising radiations and radioisotopes in the country. Part of its policy was to provide a supply of graduate physicists and chemists with radiation backgrounds for the various programmes. This included medical requirements. An early step was to encourage the University to introduce a substantial element of radiation studies in the main courses for physics degrees. One of my tasks was to introduce this section into the honours BSc course. Postgraduate training was also to be provided, including some with a medical background, with a view to producing a small cadre of physicists with specialist knowledge to satisfy the needs of the medical services. The huge Kaer el Aini Hospital in Cairo and the University Hospital in Alexandria were planning major advances in radiotherapy services, including breaking into the supervoltage field. The need for much greater physics services was being cared for well in advance. I must record my surprise at the early postgraduate lectures. I had expected a small group of recent honours graduates who had an eye on posts in the research laboratories and maybe the hospitals. In fact, for the early lectures a sizeable lecture room was filled and a large proportion of the postgraduates were far from being recent. Apparently most of the staff of the Faculty of Science were taking an interest in ionising radiations, though the interest did decline. All this was followed by a strenuous month of teaching and lecturing, by which time we could see the shape of our proposed cadre of radiation-wise physicists. It was left to the few established radiation physicists to nurture them until such time as the refurbished radiotherapy departments should need them.

The time lag before the hospitals were ready to make use of more radiation physicists was rather longer than expected. It was 1962 before my next major move. I was invited again to Cairo, this time to complete the practical training of a group of graduates who would deal with the physics of radiotherapy in the two revamped departments in Cairo and Alexandria. I was also to assist in the final commissioning of a new cobalt unit in the Kaer el Aini Hospital in Cairo and a number of other projects. This time the invitation came from the Atomic Energy Commission, which was responsible for this supply of radiation physicists.

For all this, I was established, complete with white coat, at the Kaer el Aini Hospital, and there started what was the most strenuous period of meandering of my whole career. There was a great programme of lectures, laboratory work and practical exercises, covering calibration of the new cobalt unit, isodose curves, planning of treatments, protection measurements and all that would now be called quality assurance. I was also able to make two or three brief visits to Alexandria, a fascinating experience in itself, where physics was rather better integrated into radiotherapy, though it must be admitted that most of the physical planning of individual cases was done, and well done, by the radiotherapists themselves. The deeply entrenched principle that non-medical professionals should not be directly involved with patients was difficult to overcome. In addition to this fearsome programme, I had to face a most strenuous social programme from the hospitable Egyptians. There were a surprising number of radiotherapists and physicists who had lived and trained in Britain. I quote from my travel diary: 'most of the therapists and physicists here seem to know me, even when I do not recognise them'.

This Egyptian saga continues. In 1964, as recorded in Chapter 13, when I was working out from Beirut, I was able to make another trip to Egypt and see the full set-up of radiotherapy and physics which had been tentatively planned in 1956. The end result was distinctly encouraging and this must be regarded as a modest success for the mission. This view was enhanced when, only a few weeks after I had returned home from my sabbatical with the IAEA, I returned to Egypt for another short visit. On this occasion I was to take part in a conference on uses of radioisotopes, organised by the progressive Isotope Institute of the Egyptian Atomic Energy Authority. This, I felt, showed real progress in one part of the field.

My meanderings also took me to Egypt's neighbour, the Republic of Sudan, though my effort was on a much smaller scale. Again the first contact was academic rather than professional. In 1960, I was invited to be external examiner for a rather large examination for BSc Honours Physics in the University of Khartoum. There was no specific medical involvement, though I had some radiological contacts in Khartoum. My diary records some of the horrors of running a high-level examination in the Muslim month of Ramadan, with hungry and thirsty candidates writing in temperatures of around 40 °C, to say nothing of examiners and supervisors. I made two further visits to Khartoum, one from Beirut as described in Chapter 13. They were largely concerned with advising on the equipment and protection of a substantial new department of radiology. Again it was pleasant to be recognised; most of the radiologists and at least one physicist had trained and spent time in Britain. At least I was recognisable to some as an examiner at DMR orals.

There is a small sequel to this brief story. Some time after my last professional call to Khartoum, I received an invitation from the Sudanese government to attend the opening of the new Department of Radiology. In this I was joined by Dr M Mahfous of Cairo, an old friend who had also been consulted. The formal opening was to be carried out by the President of the Republic, following a large dinner party. This was also in Ramadan and I can still visualise the President and some hundreds of guests sitting silently at tables in the open air for 10 or 15 minutes, waiting for the 6 o'clock gun to allow us to start eating. After the ceremony, Dr Mahfous and I were offered the use of a small aeroplane, with pilot, to make a two or three day tour of the cotton fields and irrigation works of central Sudan. We accepted the offer gratefully and I suggest that this is the most unusual *pourboire* ever offered to a medical physicist for services rendered. The tour was a great success, though we got very dusty in the deserts.

This story of my widening horizons even spread into my retirement years. It also involved the Arab world, though nothing very exciting is recorded. In 1970 I was asked by the World Health Organisation (WHO) to go to Tripoli in Libya to advise on and assist in the commissioning of a new department of radiotherapy. This proved to be a relatively straightforward operation. There was on site a young Libyan physicist with training in Britain and there was adequate measuring apparatus, so that it was possible to prepare the cobalt unit and other equipment for treatment purposes. There was time to go for a few days to Benghazi, where there were radiotherapy ambitions, but I did not record much success. These notes and place names may sound familiar to those with memories of the North African phases of the Second World War. I do recollect that some time after I returned home I received a gentle complaint from Tripoli that I had failed to make any recommendations on the ventilation of the treatment rooms. Such are the expectations which a hospital physicist must face.

Not long after this visit, I had a request from WHO to go to Aleppo in Northern Syria, the old crossroads for the caravan routes of the ancient world, but hardly a crossroads for medical sciences. There were plans and even some building for a new radiotherapy department, with a projected cobalt unit. The staff situation was very unsatisfactory and I was unable to offer much prospect for this proposal. Thus our survey of the wider horizons of developing medical physics ends on a rather disappointing note. I can, however, offer one interesting non-physics experience from Aleppo. The hotel in which I stayed claimed an incredible list of distinguished guests which it had entertained—politicians, soldiers, authors, even crowned heads—but I will always remember that the room allocated to me was always claimed and used by, of all people, T E Lawrence of Arabia. Could I ask for more?

In spite of this unfortunate Aleppo anti-climax, I like to think that my widespread, and maybe incoherent, meanderings may have contributed something to the birth and growth of medical physics in some of the less familiar areas of the world.

I have deliberately written at somewhat greater length on the visits to Malaysia and Egypt to illustrate two approaches to the establishment of medical physics as we now know it. In Egypt, some years before the hospitals acquired high-energy radiation equipment, the eventual need for physicists was realised. The very active Egyptian Atomic Energy Commission looked into this and selected a group of potential hospital physicists for special training. Two were sent to Britain for this training and both ended up as professors in the Cairo universities. By the time hospitals were installing cobalt and other high-energy units, relatively competent physicists were available. I was allowed, in fact, to play some part in this process.

12.9 Conclusions

The situation in Malaysia was quite different. The ambitious General Hospital in Kuala Lumpur ordered high-energy apparatus, one 25 MeV betatron and two 8 MeV linear accelerators, with little consideration for the available staff, both medical and physical. One young physicist was sent to Britain to learn the trade but he was there before any significant high-energy equipment had been established in Europe. When I arrived in the country, the machines were being installed and almost ready for commissioning. The appointed maintenance engineer arrived about the same time.

This illustrates, then, two very different approaches to the problems of setting up medical physics, at least for radiation work. In spite of these differences, I have reason to believe medical physics is now well established in both places—far be it from me to sit in judgement.

CHAPTER 13

SABBATICAL INTERLUDE

13.1 Regional Adviser: Eastern Mediterranean and the Middle East

In 1963, my meandering in medical physics took quite a dramatic turn. This event could well also have been an important point in the wider development of medical physics, at least in its radiological manifestations. By the end of the 1950s the revolution in radiotherapy was in full swing. Betatrons and linear accelerators were being installed in large numbers in established centres; indeed the linear accelerator was soon to become the standard equipment in any radiotherapy department. At the same time, telecurie units using cobalt and occasionally caesium had come into widespread use. However, the geographical distribution of these high-tech services over the regions of the world was very uneven. For example, alongside the sophistication of Western Europe, the countries of the Eastern Mediterranean and the Arab world seemed to lag far behind. This was a matter of some concern to the United Nations. This led the IAEA, with the WHO, to call meetings of experts in radiotherapy and medical physics in Vienna and then in Canada in 1962 to consider this problem. Among many proposals from Montreal was one which encouraged the IAEA to send visiting experts to appropriate developing areas, either on short-term appointment to point the way to special efforts or over the longer term to initiate and assist major developments.

One obvious candidate for this treatment was the aforementioned Eastern Mediterranean and Middle East, most countries of which clearly lagged behind in these fields. Rather cautiously, the IAEA proposed as an experiment to try a temporary (two years) appointment of a radiation physicist as Resident Regional Adviser for that area, with a view to the later development of a permanent plan. I was not personally involved in these discussions and was more than surprised when I was asked to take the post of Regional Adviser (Eastern Mediterranean and Middle East), to be based in Beirut. The original plan was for a two year appointment with vague hints of a possible permanency with the IAEA. This was a little more than I could take; one does not lightly give up a tenured Chair in a major British university. A compromise was reached when I suggested I ask for a sabbatical year from my university post, in effect just to try out the idea. To approximate to the IAEA two year plan, it was agreed that one of my junior colleagues, Tom Bryant, should go out to Beirut for two or three months to prepare the way, a task which he carried out very effectively.

One original idea was to limit the region to three or four countries, so as to allow relatively long visits by the Regional Adviser for the planning of major radiotherapy facilities and their associated physics services. As will be seen, the whole process started very slowly, but once the Adviser's office was well established, demands increased beyond expectations and eventually the programme covered some 11 independent countries. I would hesitate to suggest that this increase in demand had anything to do with the identity of the appointed Regional Adviser. Whilst not wishing to offend any of the people concerned, it is not difficult to visualise a degree of jumping on the bandwagon when there were suggestions of international assistance. Eventually, the territory of the new Regional Adviser stretched from Athens in the west to Teheran and Shiraz in the east, covering Greece, Turkey, Cyprus, Egypt, Sudan, Syria, Lebanon, Jordan, Iraq, Iran and Kuwait. The mission did not stretch to the main Gulf States which, as is well known, have their own peculiar high-level medical services. It could well be said that I knew not what I was facing and that for nearly a year, I used Beirut Airport very much as a bus station.

I still have a copy of the Adviser's job description and it makes interesting reading. The special field is defined clearly as 'Hospital Physics' and the first and, as was made very clear to me, the most important duty of the Adviser was 'To advise the governments of the participating countries as to measures required to establish or develop an effective Hospital Physics service in Radiotherapy Institutions in the countries concerned'. This rather impressive piece of job description ignored the fact that few of the countries concerned had anything remotely resembling a 'Radiotherapy Institution', though all of them

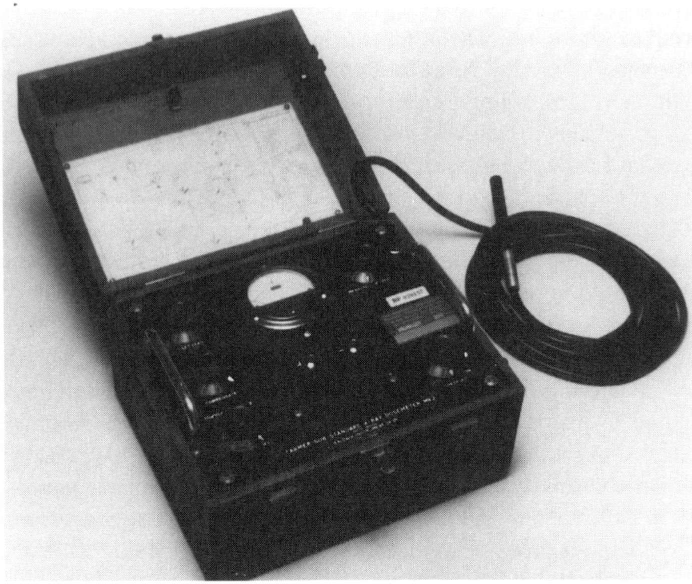

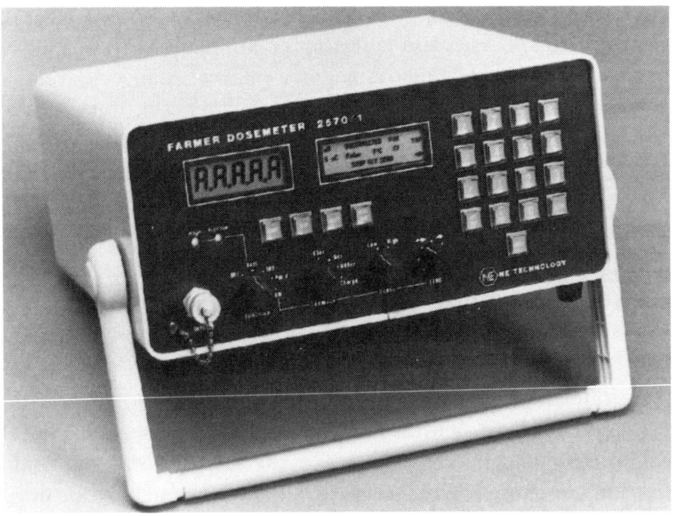

Early Farmer dosemeter (top) and secondary Farmer dosemeter (bottom).

offered some form of radiation therapy. The list of duties which followed this definition was very wide ranging, covering almost everything in radiotherapy short of the purely clinical. The clear objective was to spread the development of sound radiological physics some further 1000 miles east of Athens. Although in much of the region, radiotherapy was relatively primitive, there was clearly a hope that the clinicians would build up such services and 'Institutions' on the basis of sound applied physics. There was an interesting, slightly negative tail to the job description. Although the radiation physics of telecurie units and other sealed sources was included in my brief, 'the detailed use of unsealed radioisotopes shall be excluded', except for the designing of laboratories. Although at home our infant Institute of Nuclear Medicine was thriving, and I was still nominally its Director, nuclear medicine was out. This was the province of the WHO and there was an office at the WHO base in Alexandria to deal with it. An interesting little twist to the story arises at this point. A few days before I was due to fly to Vienna for briefing on my way to Beirut, I was asked to make a detour to Geneva to discuss these and other problems with WHO staff. In the event, relations with the WHO remained perfectly amicable and Dr Paul Taillard, their 'expert' in Alexandria, and I worked well together on at least one project in the region.

13.2 Arrival in Beirut

Thus, at the beginning of October 1963, I arrived in Beirut, just a little bewildered and very uncertain of the task ahead of me. There was one bit of encouragement. Simultaneously with my own arrival, there came a large parcel of instruments and other articles which at that time provided the normal armoury of a radiological physicist. There was a fully calibrated Farmer dosemeter, which was to become the regional standard, other dosemeters, including pocket types, two 'phantoms' and a variety of smaller items, together with some reference books (see section 4.4). This was all proof that as well as the high-level advisory and planning work, there was some practical work to be done, helping struggling hospital physics departments. In the event, this proved to be very true. The Farmer dosemeter became a familiar sight in the Customs Office at Beirut Airport, soon being passed through on the nod. At the time of writing, Beirut is just rising from the ashes of war, but in 1963 it was a beautiful, prosperous Mediterranean city which housed and entertained us for a year. It was admittedly very Europeanised, even to the extent that the taxi drivers could be addressed in English, French or Arabic. Here is a point where I could cheerfully drop another name. Most expatriate Britons and Americans arriving in Beirut tended to gravitate to the St Georges Club and I was no exception. There we received a very cordial welcome, particularly from a certain Captain Jack Mann who proved a great help in settling in recruits to the expatriate community. Many years later, this same Jack Mann achieved worldwide notice as one of the Beirut hostages.

I am not really at liberty to give a detailed account of the technical aspects of the year's work as Regional Adviser. It would undoubtedly be rather dull and, after 30 years or more, comments and criticism might well seem hurtful and unfair. Furthermore, strictly confidential reports on radiotherapy and physics planning in the various countries were sent to governments and to the Secretary General of the IAEA. Copies of these were filed in Beirut, but what happened to these is anyone's guess. I propose, therefore, for the remainder of this chapter, simply to meander through my diaries of the period, looking for items of interest for medical physics and its practitioners.

13.3 The First Two Months

The story of the first two months of the mission is one of almost unrelieved gloom and frustration. My personal diary is littered with expressions of despair at the low level of activity. One of the troubles was that, under my terms of reference, I should not take the initiative in offering service and advice to any country. The Atomic Energy Commissions or Health Departments of the participating countries were expected to initiate plans for the development of radiotherapy. Only at that stage could a request be made for the services of the Regional Adviser for advice on such things as buildings, equipment and staffing. I was learning the hard way of the style and speed of action and of some of the problems of dealing with some governments in developing countries. In self-defence, I must point out that I was not entirely idle. Lebanon was on my beat and there were two hospitals with radiotherapy departments, the American University Hospital and the Hotel Dieu, a relic of French occupation. No physicist was employed but an old student of mine was available to help. The radiotherapists presumably did their own physics, so I duly appointed myself honorary physicist to the hospitals. Using my IAEA equipment, I was able to provide a number of standard hospital physics services, such as tube calibrations, isodose curves and even a few wedge filters. Computerised treatment plans were then way into the future, but I managed an occasional hand-made plan. One small event illustrates the type of problem likely to arise. Some phosphorus-32 had been ordered for some kind of treatment applicator. When I expressed surprise at the low dose rate, I discovered that the consignment of P-32 had been held in Lebanese Customs for 10 days and in the University stores for four days before being delivered to the Hospital. I learned a lot from this work which proved a quite valuable introduction to the problems of promoting hospital physics in other countries in the region

The first sign of a breakthrough in penetrating the region came, perhaps rather surprisingly, from Damascus, almost on our doorstep. It was an invitation from the President of the University of Damascus, requesting a one day visit, largely to get to know one another and the problems. From Beirut to Damascus we usually travelled by taxi through the Lebanon–Syria border. My wife occasionally went that way for a day's shopping. The talks in Damascus were unproductive. At the time, there was clearly nothing on which to build and we had no further approach from Syria. I fear that this must be regarded as one of the failures of the mission and very disappointing. From this I learned early in my mission that the problems of developing radiotherapy and its associated physics increased as I moved east across my territory. This shows up later in the assessment of the full programme. In all, I made three visits to each of Athens and Istanbul, with one to Ankara; all the other countries requested only one or two visits in the 10 months.

13.4 Greece and Cyprus

Amidst all the frustrations of those early months, Athens eventually provided the stimulus for which I was looking. I have to admit that Greece was an easy and welcome part of my main mission. At that time, radiotherapy and radiation physics were relatively flourishing. This was partly due to the effectiveness of the Greek Atomic Energy Commission which was deeply involved in the development of telecurie treatment units. The Atomic Energy Commission (AEC) had organised a series of basic radiation physics courses at its headquarters in Democritos, providing a small but steady supply of physicists for hospital work.

Some of these also did some practical training in Britain. In all, I made three visits to Greece, in itself a great delight. The first, in November, was largely involved with the planning, together with the Greek AEC, of two rather more advanced courses in radiological physics. Quite incidentally, this visit enabled me to make a very hurried dash home to London for the wedding of our older daughter. The two training courses were held in April and September. For the April course, the IAEA invited two of my Middlesex colleagues, John Tappin and Edward Williams, as visiting lecturers. A somewhat more advanced course was carried through in September, to which the IAEA sent Dr K C Tsien and Dr M Cohen as lecturers. Both courses were extremely well received by physicists and radiotherapists and I like to think they contributed to the establishment of Greek hospital physics in the European context. As a strange tailpiece to this little history, I had to send an unusual supplementary report to Vienna. It was reported to me that there were more cobalt telecurie units either installed or planned in the Athens area than could possibly be justified by the population covered, a large proportion being installed in private clinics. I doubt whether any other country or city could outshine this.

Cyprus follows naturally after Greece in this scheme of things. At that time the political situation in Cyprus was very tense; there was open warfare with a United Nations presence. Nevertheless, I was able to make one visit with the approval of our diplomatic masters. Indeed I felt very close to the troubles in that the wife of Dimitri Souliotis, the chief radiotherapist and a very good friend of long standing, was Minister of Justice in the Makarios Cabinet. Fortunately, radiotherapy was in a quite healthy state, with a new department well on the way to operation. Tom Bryant had made a useful previous visit and we were able to make some contribution to this encouraging development in radiotherapy and hospital physics.

13.5 Turkey

Turkey comes next on the visiting list and Turkey offered a real challenge on radiotherapy and hospital physics. Illustrating this, I visited Istanbul three times and Ankara once. The first visit in December was intended as a fact-finding mission and an attempt to meet the people involved in radiotherapy and physics. This I did, but for a good deal of my time there I reverted to a conventional radiological physicist with duties of calibration of instruments and machines, producing isodose curves and so on. It was good to work in an active and reasonably well-equipped department, though I fear working hours were rather shorter than in Western Europe. One major duty was to assist in the calibration and commissioning of a large cobalt teletherapy unit. This led to a curious observation. The output of the unit appeared to be little more than half of what I would have expected. Eventually I learned that the unit was a gift from a British aid programme. The story I was told was that the unit arrived in Istanbul docks and no one had the remotest idea what it was or what to do with it. After some three years or so, a radiotherapist from the Guraba Hospital discovered the machine and the cobalt source and it found its way to its proper resting place. Turkey was fortunate again in having an active Atomic Energy Commission, which had medical physics on its agenda. The Commission, cooperating with the IAEA, made it possible to organise a major training course in radiotherapy physics in late August. Phillipe de Valence and myself from the IAEA were joined by local radiotherapists and physicists in a lecturing team and in the provision of a range of demonstrations. This course was a great success with an attendance of over 30, including some quite senior radiotherapists. It gave me at least one new experience. We had originally agreed that all the course lecturing should be in English but at the last moment the Turkish contingent had an attack of cold feet. The only immediate solution was impromptu simultaneous translation, sentence by sentence, on stage. Although, in fact, remarkably successful, this proved a tedious burden to English speakers and an appreciable lengthening of the course timetable. I like to think that our efforts with our Turkish counterparts helped in some way to bring Turkish radiotherapy and radiation physics quite a considerable way towards the mainstream of the subjects in Europe.

13.6 North Africa

The synthetic 'Region' set up by the IAEA for my Adviser mission extended south as well as east from the Eastern Mediterranean. In this way, the United Arab Republic (Egypt) and the Sudan were included. During the year I visited Cairo, Alexandria and Khartoum only once, but as I had visited these places several times on similar missions, I was able to be part of quite substantial developments. In both countries the situation concerning radiotherapy and its physics, though not ideal, was promising. Egypt, again, was fortunate in having a strong Atomic Energy Commission with a full research and development programme. The physicist situation was a little unusual. All radiological physicists were trained as postgraduates and employed by the Commission, some being sent abroad for further training. They were then allocated to hospital or other work as required.

Meandering to Alexandria, a pleasant experience in itself, gave an opportunity to link up with the WHO Regional Office there and particularly with Paul Taillard, covering radiology and nuclear medicine with the WHO. In fact Taillard joined me on a visit to Khartoum where we were both involved in advising on a new department of radiology and nuclear medicine. For a developing country, the physics side of the programme was well covered, one or more physicists having been sent to Britain for training well before the completion of the building programme.

13.7 East from Beirut

The remainder of our region, east of Beirut, presented quite a number of different problems. It was made up, of course, of mainly Arab and Muslim countries, with a different way of life, different living speed, different work ethic from those to which we are accustomed in Western Europe. The medical practitioner there demands and receives a very high degree of authority and respect; this does not simplify the promotion of a parallel profession of medical physicist. In this area, I made one or more visits to Jordan, Syria, Iraq, Iran and a brief emergency visit to Kuwait. Syria, and my lack of success there, have already been mentioned. Some years after my return to Britain, I was invited by the WHO to go to Aleppo, in Northern Syria, to assess the possibilities of a cobalt telecurie treatment project there. As Chapter 12 shows, the result was equally disappointing. Iraq and visits to Baghdad presented something of an enigma. In Baghdad, there were elaborate plans for a major cancer treatment centre and building was proceeding. I clearly remember, in a day temperature of 45 °C, watching the pouring of hand-mixed concrete for the walls of a proposed cobalt unit treatment room. I even remember, with some feeling, condemning some of the walls where the concrete had not been properly vibrated. I was not popular. I had to express great concern about the staffing situation.

There were visions of a fine new treatment centre two years hence, with little or no specialist staff to run it. Radiotherapists were very thin on the ground and hospital physicists non-existent.

Moving east again to Iran produced a different and perhaps an even more complex situation; here were signs of European and even British influence. In Teheran there was a limited amount of radiotherapy, including two small cobalt units, one private. There were two competent hospital physicists, though, in the strange employment situation in Iran, each of them had to have two professional appointments in order to make a living. Mention of the physicists brings out another unrelated reminiscence. A small party consisting of five physicists, including myself, two electricians and a radiologist was looking round the Cancer Centre and came to inspect the cobalt unit suite. We found no lighting and no electric power and searched long for the fault. Just as we were about to leave in despair, a little old cleaning lady, dressed in the native black, walked in, saw our problem, went to the appropriate cupboard, switched on the power and left. Not a word was spoken. I must confess that at that moment I was not optimistic for the progress of radiotherapy and physics in that part of the world.

Being Teheran in 1964, any visit of this kind involved a little more than medical physics. Brian Windeyer and Stanford Cade were there on some very high-powered medical diplomatic mission. This involved some extra meetings and receptions. It also took me very close to the gaffe of all time in medical physics history. One morning I was due to do some work at the Cancer Centre, whilst Brian and Sir Stanford were due for an audience with the Shah. Yes, Iran (Persia) had a Shah in 1964. Official cars arrived to transport us and off I went to work. Ten minutes later, even with my limited knowledge of the streets of Teheran, I realised we had taken a wrong turn and were heading for the Royal Palace rather than the Cancer Centre. Obviously I was in the wrong car and there was panic. Fortunately we got back to the hotel in time and the diplomatic doctors just made it in time for their audience.

Iran also provided me with a fascinating side trip to more romantic scenes of ancient Persia, the lands of Darius and Xerxes and the palaces of Persepolis. I travelled south by way of Isfahan to Shiraz, where I had been invited to advise on the position of radiotherapy and physics.

Largely as a result of American influence, the situation in Shiraz was quite promising and later, just as I was finishing my stint as Regional Adviser, a substantial development programme was submitted. Perhaps this was one of our successes.

I had to see something of ancient Persia and it was agreed that a visit to the ancient palaces of Persepolis was just possible. This involved very early rising, suffering from a typical Eastern digestive disorder, and lecturing at 7.30 am to the combined hospital staffs. Nevertheless, such a visit provides a talking point for a lifetime, even though it has little to do with medical physics.

There remain just the two smaller Arab countries on my list, Jordan and Kuwait. In Amman, Jordan, a very fine new department of radiology had been built under very difficult conditions. This was quite outside my remit, even though I was impressed. Some time later I heard from the IAEA in Vienna that it had received a bitter complaint from the Amman radiologist that the IAEA 'expert', in his report, had completely ignored his beautiful brainchild. Fortunately, Vienna understood. Unfortunately, there was no serious development of radiotherapy and there was no sign of a hospital physicist. From the point of view of my mission, Jordan was, for the time being, not a good prospect.

My only visit to Kuwait was in answer to an emergency call, which was not strictly my *raison d'être* in Beirut. The Kuwait national secondary standard dosemeter had developed a fault. Kuwait had a very adequate radiotherapy service, well staffed and with a very competent British physicist on site. With the regional standard Farmer dosemeter and other equipment which I carried round with me, the main dosimetric problems were soon solved. From Kuwait there come clearer memories of the 'perks' which occasionally appear for an official visitor with a United Nations passport. It happened that I arrived in Kuwait after midnight, with no warning, and had to fend for myself. With the help of a taxi driver I ended up in a rather inelegant small Arab hotel. However, when the Ministry of Health learned of my situation next morning, I was quickly whisked away to a luxury suite in an hotel of at least five stars which had refused me the night before. Outside was a huge Cadillac with a white-robed chauffeur, at my disposal at all times. Who would be a hospital physicist?

13.8 Conclusions

The above paragraphs have been written quite deliberately and perhaps a little light-heartedly as a simple account of the existing state of affairs in medical physics and radiotherapy at that particular time and in that rather artificial region. I have not attempted to assess the impact of my own IAEA presence on developments. This would clearly be a long-term operation. Furthermore, any attempt at a full assessment of the subsequent 30 years would be way beyond my resources. I suppose it could be regarded as a very small part of the history of hospital physics in one part of the world. A perfectly legitimate question to be asked would be: did the IAEA idea of a Regional Adviser work? To be perfectly honest, from the point of view of the original job description, it was not a resounding success. There was no great surge of hospital physics or of high-energy radiotherapy in the Middle East, as there was in Western Europe and the United States in the 1950s. So far as I know, that strange divide at the Levant coast still exists. There is, to the best of my knowledge, no regional professional organisation for medical physics in that area, although such bodies flourish in most parts of the world.

In spite of this pessimistic judgement, I believe fruitful seeds of medical physics were sown. Khartoum got its fine new radiological department with its physics. Turkey, Greece and Egypt have all made significant progress towards becoming part of the worldwide family of medical physics. Further east the story is far less happy. In the last 20 or 30 years, political and cultural developments, combined with intense conflict in some places, have in no way been conducive to technical progress. For any seed sown there, we can only hope. For myself, that sabbatical year in Eastern Europe and Western Asia was a fascinating, almost shattering, experience, even though, with hindsight, some of it could only be described as 'mission impossible'. To produce major developments in radiotherapy and its associated physics in one year in 11 countries is very clearly an impossible task for one man. It may have worked for two or even three countries, but for 11, no. Even so, I still believe it was a worthwhile exercise.

Considering the rather poor report on medical physics, an addition, almost an apology, seems necessary. In a recent *Bulletin of the International Organisation of Medical Physics*, I find a report of a course/workshop on radiation therapy physics held in September 1996 in Rabat, Morocco, organised by the Moroccan Organisation for Medical Physics. Now, in 1964, Morocco could hardly be considered on the medical physics map, although there was a spark of light in nearby Libya. It seems possible that my pessimistic thoughts about medical physics in the Middle East and the Arab world could have been premature. If that were the case, then no one would be more pleased than myself.

CHAPTER 14

MEDICAL PHYSICS HAS ARRIVED

14.1 The Profession of Physics Applied to Medicine

I have tried to make clear throughout these meanderings that I had no desire to attempt a definitive history of medical physics. A preliminary attempt at this was well done for the 40th anniversary celebration of the HPA and any major expansion would require resources beyond my range. I repeat this statement here in case the intentions of this chapter are questioned. During these meanderings, I have observed, consciously or unconsciously, a whole range of changes in the development and acceptance of physics as a fully recognised member of that group of professions which must surround the practice of medicine. In 60 years or more there have been subtle changes whereby the physicist has become an integral part of the clinical team. In this matter, I suppose I should consider myself fortunate in being able to act as a fairly unbiased observer. Throughout most of my working life I have had the advantage of the professional status of a university teacher, whatever, at any particular time, my hospital relationship may have been. Even this situation was sometimes slightly ambiguous. Although there had been a number of professors of physics in the medical schools since the turn of the century, there were unspoken queries about their academic status in mainstream physics. Like it or not, physics applied to medicine had not quite made the grade academically. Nevertheless, from the security of even that somewhat doubting environment, it was possible to observe the rather erratic rise in the professional credibility of medical physics.

If we go back to my own entry into the field in 1932, there was little or nothing which could be described as a profession; there was just a small group of individuals, widely separated, facing the challenge of the measurement of radiation dose in a form acceptable to the radiotherapists. There were side interests but X- and gamma-ray dosimetry was the key to progress. One curiosity naturally intrigued me. The *Medical Directory* showed Sidney Russ as Consultant (known before the National Health Service as Honorary) Physicist at the Middlesex Hospital.

At that time a major factor in the support of what was to become medical physics was the BIR. Although having a predominantly medical membership, even from its inception, the Institute has always regarded medical and non-medical members as of equal merit. The Presidents have always been medical and non-medical in alternate years; indeed the very first President in 1897 was a physicist. There was, therefore, in the 1930s a small group of physicists from the hospitals and medical schools, the National Physical Laboratory and elsewhere who were trying to build the scientific foundations of radiology and radiotherapy. As we have seen, during the 1930s, there was a small but steady influx of individuals into the field, some of whom were to claim fame in later years. The time, however, was not quite ripe for professional organisation; this phase was described in some detail in Chapter 5 and will not be repeated here.

14.2 The National Radium Commission and the Founding of the HPA

In this context, the year 1937 looms as an historical landmark. As mentioned in Chapter 5, in that year the National Radium Commission insisted that any department of radiotherapy which used radium must employ a properly qualified physicist. It might well be said that at this point, the profession of radiological physicist was born, perhaps as an embryo form of our modern medical physicist. The result was, as we know, a steady increase in the number of physicists as new radium centres and radiotherapy departments were designated.

This period, from 1937 to the outbreak of war, though based mainly on an enthusiasm for the use of radium, was vital in the emergence of medical physics as a profession. It led, of course, to what was the most important point to date in that history, the founding of the HPA in 1943. Sidney Russ, who was deeply involved himself in the Radium Commission, saw clearly that if the physicists were to be their own men (or women!), some kind of organisation was necessary. Even from the beginning there were problems in the form the organisation should take and exactly the people who should be involved. I understand that Russ's original idea was that at least the earlier meetings should be limited to heads of hospital physics departments. Apparently this idea was most unpopular and was quickly thrown out. A number of hospital physicists in the north of England had already formed an association which would eventually become the Northern Branch of the HPA and they had quite a clear idea of what they wanted. The mode of action and field of activities of the new Association also caused problems. A number of people, including Russ, were quite certain that the HPA should be purely and simply a scientific society concerned only with the technical and scientific applications of physical science to medicine. Such things as conditions of service for physicists in the National Health Service, salaries and other financial matters were objectionable and should not be considered. They were the province of trade unions and, at that time, trade unions were not very popular in some quarters. Nevertheless, almost certainly affected by the infant Northern Branch, the wider interpretation of the objectives was accepted. The HPA came to life as a combination of scientific society and trade union. There were, in the beginning, some doubts about the viability of the organisation, but sure enough, with a hitch or two, it served the burgeoning profession bravely for almost exactly 50 years.

Writing these notes, with hindsight, about the birth of the HPA brings to mind some ideas and thoughts from that period which, though strongly felt at the time, were rarely expressed openly. These concerned the position and status of the physical scientists as regarded by the hierarchy, professional and administrative. In some places, particularly those with university attachment, there was no great problem; the physicists were accepted as professional graduates on much the same terms as their counterparts with medical qualifications. Sadly, this was not always the case as new radium centres and radiotherapy departments were set up and the compulsory physicist became part of the scene. In some places there was a tendency to regard this member of staff as a slightly superior type of auxiliary worker, to be given orders rather than to be consulted on the problems of radiotherapy. I admit this may sound like something of an exaggeration but I have no doubt that it came quite close to the situation which existed in many developing centres.

It was in these circumstances that I believe the HPA made its major contribution to the establishment of the profession. On the 'learned society' side of its activities, though they were not extensive, the Association maintained the highest standards in scientific meetings and publishing as well as preparing the way for establishing acceptable specialist qualifications in the field. At the same time, members concerned with such sordid matters as pay and conditions of service fought and won their battles in Whitley Councils and elsewhere for the full recognition of the hospital physicists. When the time was ripe, the HPA was ready to metamorphose into the IPSM and eventually the Institute of Physics and Engineering in Medicine, to take its rightful place among the learned societies in the field.

14.3 Consultant Adviser (to the Chief Medical Officer)

In 1961 I believe there was a significant turning point in the journey to full professional recognition, although I take it up with some diffidence as I was unwittingly personally involved. The technical and professional organisation of the Ministry of Health, and maybe of other ministries, is something of a mystery to most people, not least to those who work in the National Health Service. Not very well known at that time in that Service was the Consultant Advisers Group, some 30 or 40 consultants, at least one from each speciality and two or three from the bigger ones. They were appointed to advise the Chief Medical Officer, and through him the Minister, on major day-to-day problems which arise in the Service. They met regularly to discuss and advise on more general problems of clinical and public interest. Some time in 1961, I know not how, someone, I know not who, observed that physics was playing an increasingly important part in some major clinical situations. It was decided, I know not where, that a medical physicist should join the exalted ranks of the Consultant Advisers. Knowing nothing of this structure, to my intense astonishment, I was invited to become a member of these very ranks. I was even offered a handy honorarium for this service. There I sat, then, a lone physicist with the assembled cream of medical and surgical consultants sorting out the National Health Service. It was quite an experience. Besides these meetings, the appointment involved advising on day-to-day physics problems from hospitals and authorities which arrived on the administrators' desks at the Elephant and Castle. A small amount of visiting of physics departments was involved though if I was called to Scotland, the permission of the Scottish Home Department was required; it was never refused. This appointment lasted until some short time after my retirement from the Middlesex, when Jack Boag took over. I would not presume to suggest that this move had significant effects on the working lives of hospital physicists of the time. Nevertheless, some ice was broken and I believe that from that time onwards, the true professional status of physicists in the National Health Service was established. The new appointment even involved making an occasional nomination for the Honours List. I have an idea that this advisory board no longer exists in this form, its duties having been taken over by nominees of the Royal Colleges and the Conference of Clinical Sciences.

14.4 The Zuckerman Committee

The increasing impact of physics services on sections of clinical medicine other than radiotherapy was being matched by other branches of science and technology. Apparatus for biochemical analysis had become largely automated and increasing amounts of high-tech apparatus were appearing in operating theatres, wards, intensive care units and so on. Inevitably this involved increasing numbers of graduate scientific and engineering staff and of high-level technical assistants. There was clearly a case for a substantial rethink of the place of science and engineering in the hospital service. Thus we had in 1967–1968 the government-appointed Committee on Hospital Scientific and Technical Services, the famous (or infamous according to point of view) Zuckerman Committee. Rather hesitantly I accepted the invitation to be a member of that committee with the obvious purpose of representing medical physics from the inside. My secretary, at least, found pleasure in my membership. On the days when the Committee met, she delighted in informing callers that the Professor was not available; he was at the Cabinet Office. The simple setting up of the Committee aroused a variety of emotions. There were many welcoming noises but also a whole range of fears and objections. There were fears that graduate scientists, without the magic 'MB' qualification, would take on clinical responsibilities. There were ideas that surgeons and physicians, with only a fragile scientific base, would take charge of science and engineering departments and supervise the service and research work of highly qualified scientists. Highly skilled technicians in a variety of disciplines feared they might be torn from their beloved equipment and techniques and deposited in a central pool and so on. There was, in fact, never any real risk of any of these disasters occurring.

No useful purpose would be served by discussing the report of the Zuckerman Committee in detail here; it is, indeed, ancient history. In long-term retrospect, it is clear that some of those recommendations were never fully implemented in detail. At the same time, most of the purposes for which the Committee was set up were served. Most obvious of the positive developments has been the steady growth of independent physics and engineering departments, working under a variety of rather exotic names and serving, as required, most of the clinical departments. Some independent scientists in clinical departments still exist, in spite of the recommendations of the Committee. At the same time, the auxiliary professions, if such a description is permitted, the radiographers, the physiotherapists, the laboratory technicians and others who feared, under the Committee, for their independence, quite rightly stayed and prospered with their parent departments. The whole pattern gradually sorted itself out to the satisfaction of most people. In a few enlightened organisations, according to their advertisements for staff, the term 'Consultant Physicist' has even been heard—shades of Sidney Russ.

14.5 The IPEM

Following all this excitement, it might truly be said that medical physics entered the final stages of the process of becoming a totally independent profession. This is most clearly demonstrated in the emergence of the IPSM and the Institute of Physics and Engineering in Medicine. For over 50 years, the HPA had filled the dual role of caring for and negotiating the conditions of service of physicists in the National Health Service and also for the scientific and professional standards of that work. The time had come to separate these activities. Conditions of service affairs were devolved without great difficulty elsewhere, leaving behind all the aspects and trappings of a professional learned society. The success of the Institute has been quite dramatic over its relatively short life. It is now a strong, clearly identified organisation, with a highly respected leadership, within the professional hierarchy of medicine. The Institute has an outstanding publications service, linked with The Institute of Physics, and a large number of committees and specialist groups, covering almost every conceivable application of physics and engineering to medicine.

At a late stage in writing, there is a brief supplement which must be added to this story. The year 1995 saw a most important tidying-up operation. During much of its lifetime the Biological Engineering Society (BES) has grown up alongside the HPA and the IPSM. The ultimate objectives of the two societies were similar, or more strictly complementary. There has been over many years a reasonably good-natured rivalry. So long as the majority of hospital physicists were involved mainly in the problems of radiation and radiotherapy, this division was unimportant, but as both groups began to diversify, their *raison d'être* as separate bodies became less relevant. Eventually, the IPSM and the BES have become one learned society, to be known as The Institute of Physics and Engineering in Medicine (IPEM). This merger has been finalised as these notes are being written and the new body is sorting out its many problems. It would be very improper to make serious comment on the detailed situation at this stage. Suffice it to say that the omens for the future of the profession with its new representative body are promising. To help my promise to avoid definitive history, may I suggest that this chapter should be read, like all the others, as the mental meanderings of an interested party amongst the tangled affairs of a growing scientific organisation?

14.6 Conclusions

We started this pseudo-historical essay on medical physics as a profession with half a dozen young graduate physicists rather eccentrically practising their tricks with radiation in hospitals rather than the more usual university laboratories. We finish with a brand new and burgeoning profession, with many thousands of members worldwide and, in this country, a brand new and widely accepted learned society, taking its place in the body of science, engineering and medicine. This is not a bad effort to be observed in one lifetime.

CHAPTER 15

SITTING DOWN

Perhaps the rather unexpected title for this chapter should be explained without delay. Through my lifetime, in addition to my meanderings in medical physics, I have been a loyal (I hope) member of the Methodist Church. When a minister of that church decides to retire from active work, he applies for the privilege of 'sitting down'; this quaint phrase even appears in the official documentation. As my professional activities were winding down and my meanderings coming to an end, I felt that 'sitting down' was an appropriate description of my situation. I had meandered long, hard and far, and maybe should now relax.

15.1 Retirement

The time was the end of the 1960s and although much was still happening in the forefront of medical physics, the enormous surge of development which started just after the war ended was maybe slowing down a little. My own Department, with its teaching, research and hospital services, had survived well the meanderings of the previous 20 years or so. With a tenured University Chair, my personal position was secure enough, though I had distinct doubts about the way ahead. Admittedly the doubts were vague but I was not entirely happy about the way in which the National Health Service was moving. There were also anxieties, later to prove real, about the pattern of medical schools in London. Inevitably questions arose in my mind as to whether I could or even whether I wanted to cope with these problems ahead. I admit I inclined to the idea that major developments in science or administration were rarely initiated by men or women over 60, though I could be shot down for this. Already I had around me a very fine department, a department of which I was inordinately proud. None of this, of course, made the decision any easier. However, after much heart searching, I decided in 1969 to 'sit down', retiring at 62, the earliest age permitted, and not claiming the further five years normally allowed by the University.

Although I was not directly involved, the search for my successor was most interesting. Eventually, to my great pleasure and satisfaction, the post was offered to and accepted by James F Tait, FRS. James was, of course, one of my cherished 1950s group of up-and-coming medical physicists at the Middlesex, who have made such widespread contributions to the subject. He and his wife, Sylvia, had, in the intervening years, made large and important contributions to the study of adrenal steroids in the United States. The physics of endocrinology, a branch of the subject not particularly well known to most medical physicists, was coming back home to our Middlesex Physics Department. After the excitement of the appointment came the problems. First of all, the Taits' commitments in the United States meant that they could not come until 1970. There were suggestions of reversing my retirement but it was too late. As a compromise I took my formal retirement and agreed to return to the post on a half-time basis for one session. This meant commuting from Norfolk three times a week but it worked out remarkably well. This became part of a gradual retirement process spread over about three years, a process which can be strongly recommended. There is a great temptation to concentrate on cultivating the garden or on travelling the world. I retained several examinerships for two or three years which meant some travelling around the country and contacts with physics teachers. I also stayed on a number of committees of the Ministry of Health and the Radioactive Substances Advisory Committee, whilst retaining my position as Consultant Adviser at the Ministry of Health. One intriguing committee which I chaired was the Miscellaneous Sources Panel of the Radioactive Substances Advisory Committee. This committee was responsible for advising on the various instruments, tools and gadgets used in everyday life which could include small radioactive sources in their make-up, obvious examples being watches, luminous signs and so on. Many people are sorely puzzled when they are told how their cabin night sign or their night fishing float are made luminous. All this meant that for a time, each month, there would be one or more excursions to London, Coventry or elsewhere, not a strenuous life but one with some added purpose. As recorded elsewhere, I made two longer excursions on behalf of the WHO to Libya and Syria.

As with many other chapters, there has to be an epilogue, perhaps, in this case, two. After the run-down of my active meandering, I was kept in touch with medical physics largely through my membership of the BIR, the HPA and later the IPSM. It is a fortunate fact that, being a one-time editor of the *British Journal of Radiology* and the Founder Editor of *Physics in Medicine and Biology*, I receive regular copies of these distinguished journals. The least I could do was to read them, though I soon began to feel it hard going. Nowadays I have to admit that much of what I find in the journals is way beyond my comprehension. This can be a great frustration, but to one who started very near the beginnings it is a matter of great satisfaction.

15.2 Staff Mascot

It just happens that my travels in medical physics did not quite end at this point. After some 18 years of retirement in Norfolk and attaining very advanced years, we moved to Tewkesbury in Gloucestershire to be near family. This almost started up a new meandering. I had memories of coming to this part of the world many years ago and putting on a very superior smile at the public appeals for 'a cobalt unit for Cheltenham'. Hadn't I recently introduced the first of these machines to Britain? My new home lay roughly midway between Gloucester and Cheltenham, each of which has a fine hospital, Gloucester Royal and Cheltenham General. At one time they each had a department of physics but now they are one, a single department under the care of Professor Angela Newing. How Angela copes I will never know, but it seems to work. News of our arrival in Tewkesbury reached the hospitals and I soon became some sort of honorary member of staff (perhaps mascot would be a better word), receiving invitations to staff meetings and seminars, held every few weeks. On one occasion I was even allowed to address the seminar. It was indeed a minor revival of the meandering, though I cannot now hope to make any technical contribution.

15.3 A 90th Birthday Seminar

There was, indeed, a very appropriate ending to my meanderings. At the time of my 90th birthday, Angela Newing and a number of friends organised a seminar to celebrate the occasion; I have to admit that I have always thought of this as a Teutonic method of celebrating the

Professor J E Roberts on the occasion of his 90th Birthday Seminar.

professor's birthday. The seminar, at Frenchay Hospital in Bristol, was a great success. Around 80 people from all parts of Britain were there; it was a real joy to welcome Frank Farmer, almost as old as myself, who had travelled from Newcastle to Bristol for the occasion.

Two of the old-timers, Len Lamerton and Alan Jennings, talked of the problems of radiation measurement which plagued our earlier days. Peter Wells, recent President of the IPEM, looked at the question 'Is your physics really necessary?' rather more seriously than I did when it formed the title of my Presidential Address to the BIR. Edward Grant told of the worldwide expansion of the dielectric studies of biological materials which started as a PhD thesis at the Middlesex. Two from the present generation, Dr W Vennart, whose father was one of our pioneers, and Mr M Casebow, brought us right into the 1990s. For me it was an unexpected experience out of this world. My meanderings, which at times must have seemed rather wayward, had landed me right back in the heart of medical physics. What more could I ask?

INDEX